יְהִי רָצוֹן מִלְּפָנֶיךָ ה' אֱלֹקֵינוּ וֵאלֹקֵי אֲבוֹתֵינוּ, שֶׁתְּבָרֵךְ כָּל אִילָנוֹת הָאֶתְרוֹג לְהוֹצִיא פֵּירוֹתֵיהֶם בְּעִתָּם, וְיוֹצִיאוּ אֶתְרוֹגִים טוֹבִים יָפִים וּמְהֻדָּרִים וּנְקִיִּים מִכָּל מוּם, וְלֹא יַעֲלֶה בָּהֶם שׁוּם חֲזָזִית, וְיִהְיוּ שְׁלֵמִים וְלֹא יִהְיֶה בָּהֶם שׁוּם חִסָּרוֹן וַאֲפִילוּ עֲקִיצַת קוֹץ וְיִהְיוּ מְצוּיִים לָנוּ וּלְכָל יִשְׂרָאֵל בְּכָל מָקוֹם שֶׁהֵם, לְקַיֵּם בָּהֶם מִצְוַת נְטִילָה עִם הַלּוּלָב בְּחַג הַסֻּכּוֹת שֶׁיָּבֹא עָלֵינוּ לְחַיִּים טוֹבִים וּלְשָׁלוֹם, כַּאֲשֶׁר צִוִּיתָנוּ בְּתוֹרָתְךָ עַל יְדֵי מֹשֶׁה עַבְדֶּךָ (ונ"א עבדך), וּלְקַחְתֶּם לָכֶם בַּיּוֹם הָרִאשׁוֹן פְּרִי עֵץ הָדָר כַּפּוֹת תְּמָרִים וַעֲנַף עֵץ עָבוֹת וְעַרְבֵי נָחַל, וִיהִי רָצוֹן מִלְּפָנֶיךָ ה' אֱלֹקֵינוּ וֵאלֹקֵי אֲבוֹתֵינוּ, שֶׁתְּעַזְרֵנוּ וְתַסִּיעֵנוּ לְקַיֵּם מִצְוָה זוֹ שֶׁל נְטִילַת לוּלָב הֲדַס עֲרָבָה וְאֶתְרוֹג, כְּתִקְנָהּ בִּזְמַנָּהּ בְּחַג הַסֻּכּוֹת שֶׁיָּבֹא עָלֵינוּ לְחַיִּים טוֹבִים וּלְשָׁלוֹם בְּשִׂמְחָה וּבְטוּב לֵבָב, וְתַזְמִין לָנוּ אֶתְרוֹג יָפֶה וּמְהֻדָּר וְנָקִי וְשָׁלֵם וְכָשֵׁר כְּהִלְכָתוֹ.

The Four Minim

A Practical Illustrated Guide

Rav Avraham Chaim Adess

ארבעת המינים למהדרין

The Four Minim

A Practical Illustrated Guide

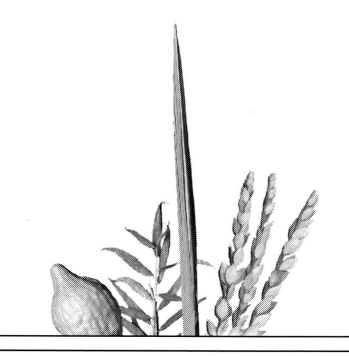

Translated by Daniel Worenklein
Edited by Rabbi David Kahn

Originally published in 2003 in Hebrew as *Arba'as HaMinim LaMehadrin*
Copyright © 2003 Hebrew edition by Rabbi Avraham Chaim Adess

Copyright © 2004 English edition by Feldheim Publishers

ISBN 1-58330-724-9

All rights reserved.
No part of this publication may be
translated, reproduced, stored in a retrieval
system or transmitted, in any form or by any
means, electronic, mechanical, photocopying,
recording or otherwise, without prior permission in
writing from the publishers.

FELDHEIM PUBLISHERS
POB 43163
Jerusalem, Israel

208 Airport Executive Park
Nanuet, NY 10954

www.feldheim.com

Printed in Israel

התודה והברכה

נתונה למורינו ורבינו

מרן הגאון ר' חיים קנייבסקי שליט"א

על אשר עוררני להוציא קונטרס זה,

ואמר לי שיהא תועלת רבה לילדי ישראל

"ברכה והצלחה"

אביע תודה מיוחדת

למורינו ורבותינו גדולי ישראל

מרן הגאון ר' חיים קנייבסקי שליט"א

ומרן הגאון ר' ניסים קרליץ שליט"א

המאירים דרכינו בתורת יקותיאל

אורים ותומים לכל שואל

ובהנחייתם ועצתם את חיבורי זה נחלני אל

יה"ר שימשיכו להנהיג מתוך בריאות איתנה ואת העם לנהל

בדרך ובמסילה המובילה בית אל

עד ביאת גואל בהשראת שכינה בבית אריאל

הסכמת מורנו ורבנו מרן ראש הישיבה
הגאון רבי יהודה עדס שליט"א
ראש ישיבת קול יעקב

בס"ד ירושלים עיה"ק תובב"א, י"ז סיון תשנ"ז

הנה הרה"ג הנעלה, אברך מיוחד ומעולה הרב אברהם חיים עדס שליט"א בן אחי אהובי הרה"ג כמוהר"ר רבי דניאל עדס שליט"א.

כבר פקע שמיה בספרים החשובים שיצאו מתחת ידו ספרי ויאמר אברהם בעניני או"ח, ועתה הניף ידו בעניני יו"ד, וכמעשהו בראשונה כן עתה הוציא מתחת ידו דבר נאה ומתוקן בשום שכל וסברא ישרה, וכל דבריו נאמרים בטעם רב. ובמעט שעברתי בין בתריו אמינא שכדאים הדברים לעלות ע"ג ספר ולהיותו זוכה זוכה להקריב קרבנו להיות עולה על שולחן מלכים, מאן מלכי רבנן.

וברכתי תפילתי להמשיך כהנה וכהנה בעמל התורה ולזכות להמשיך את תפארת המשפחה בשלשלתא דדהבא של תורה היקרה מפז ופנינים.

כן דברי דודך אוהבך

יהודה עדס

ברכת שרי התורה
גאוני בד"ץ העדה החרדית
פעיה"ק ירושלים ת"ו

יום כ' לחדש סיון תשנ"ז

הנה כבר איתמחי גברא בחלקים הקודמים של ספרו הנכבד "ויאמר אברהם" שנתקבלו בחיבת הקודש ה"ה הרה"ג המצויין ומופלג בתורה ויר"ש בן גדולים כש"ת מוהר"ר אברהם חיים עדס שליט"א.

וכעת מדפיס אידרא חלק ששי, וכמעשהו בראשונים כך מעשהו כעת בחו"ב באר היטב.

ואם אמנם אין מדרכנו לתת הסכמה על ספרי הלכה אבל מכיון שיש תועלת בהדפסת ספרו וזיכוי הרבים, הננו מברכים אותו שיצליח להגדיל תורה ולהאדירה, ולזכות את הרבים בעוד ספרים נכבדים, ויבואו אחינו בנ"י ויביאו את הברכה לביתם בכסף מלא ויהנו ממנו.

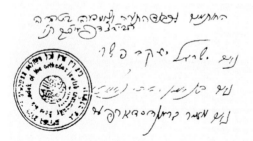

הסכמת הגאון מעמודי ההוראה רבי משה הלברשטאם שליט"א
מו"ץ בעדה החרדית ומח"ס "דברי משה"

בס"ד

היה למראה עיני ספרו היקר והנעלה של מע"כ ידידי האברך הה"ג מופלג בתוי"ש ובנן של גדולים מוהר"ר אברהם עדס שליט"א מפעיהקת"ו, אשר בשם ויאמר אברהם יכונה, ובו העלה הלכות ברורות בהלכות ארבעת המינים, וסביב ההלכות הרחיב את הדיבור בבירורי ההלכות בבקיאות ובסברות ישרות ונכוחות, והן אמנם אשר מפאת היותי עמוס בעבוה"ק לא עלה בידי לעיין כראוי בכל הספר ובפרט באשר הדברים נוגעים להלכה, אך כבר בהשקפה לטובה נוכחתי לדעת כי הולך מישרים צולל במצולות דברי הפוסקים ומעלה פנינים יקרים, ובאשר ידענא ליה מכבר הימים לת"ח מופלג, וגם כבר חיבר והו"ל ספרים נכבדים בבירורי סוגיות הש"ס, וראיתי את אשר הרבה בשבחו ידי"נ הגאון, מזה בן מזה, ומרביץ תורה ויראה לעדרים, כש"ת רבי יהודה עדס שליט"א ראש ישיבת קול יעקב, פעיה"ק ירושת"ו, בן לאותו גאון וצדיק חו"פ המפורסם רבי יעקב עדס זצוקלל"ה.

אשר על כן אף ידי תיכון עמו לתמכו ולחזקו שיזכה להוציא מחברתו לאורה לזיכוי הרבים, שיהנו מנועם דבריו הנאמרים בטוב טעם ודעת, וזכות אבותיו תהא מסייעתו שיזכה לילך מחיל אל חיל ולעלות במעלות התורה והיראה ולהפיץ מעינותיו חוצה מתוך שמחה ונחת וכטו"ס אכי"ר.

מכתב ברכה ממרן המשגיח הגה"צ רבי דב יפה שליט"א
משגיח בישיבות כפר חסידים וקול יעקב

בס"ד מנ"א תשנ"ד

לכב' ידידי היקר והאהוב הרבה הדבוק בעיון התורה ויראת ה' היא אוצרו הרה"ג אברהם חיים עדס נ"י בן ידידי היקר והדגול המסור מאד להעמדת הישיבה המיוחדת במעלותיה ישיבת קול יעקב הרה"ג רבי דניאל עדס שליט"א.

אחדשה"ט וש"ת בידידות רבה

שמחתי לראות כי דבקותך בעיון התורה ללון בעומקה של הלכה בצירוף יראת השי"ת ומדות נאות הניבו פירות נאים משמחי לב, חיבור חשוב על עניני ארבעה מינים כפי שהעידו גדולי תורה.

ויהי רצון שיהא חבורך להגדיל תורה ותלך מחיל אל חיל לחבר עוד חיבורים חשובים.

ברכת אוהבך מכבדך

דב יפה

מכתב תהלה ממרן הגאון רבי חיים פנחס שיינברג שליט"א
ראש ישיבת תורה אור ומורה הוראה דקרית מטרסדורף

בס"ד ט' אד"א תשנ"ה

לכבוד האברך היקר הרה"ג אברהם חיים עדס שליט"א בנן של גדולים.

שמחתי לראות את ספרך ויאמר אברהם על הלכות בורר שנעשה בעמל ובכשרון רב, ואמינא שכדאים הדברים לעלות על שולחן מלכים מאן מלכי רבנן.

ומיוחד יש בספר זה שהשו"ע והמשנ"ב ושאר פוסקים מפורטים לפרטים.

על כן ברכתי למחבר שליט"א שיפוצו מעינותיו חוצה ויזכה להגדיל תורה ולהאדירה.

בברכת התורה

חיים פנחס שיינברג

הסכמת הגאון הגדול רבי אפרים גרינבלאט שליט"א
רב ושו"ב במעמפיס מח"ס שו"ת רבבות אפרים ז"ח

בס"ד י"ז סיון תשנ"ז

עברתי בין ערוגות הבושם של הספרים "ויאמר אברהם" ד' חלקים, אשר חברו ידידי החשוב, האי גברא יקירא הרה"ג, חריף ובקי דנחית לעומקא דדינא, הרב אברהם חיים עדס שליט"א, וראיתי בספרים חידושים נפלאים בכל מקצועות התורה, וראיתי שהוא גברא רבה ואדם גדול בתורה, ועומד כעת להוציא לאור ספר ויאמר אברהם על הלכות תערובת, וביקש הסכמתי, ורצונו של אדם זהו כבודו, והנני נותן לו הסכמה מלאה, ולהכירו במקומות שלא נודע, כגאון עצום, וישיקחו ספריו וספר החדש אל ביתם, ויהנו מהספר, ואני מצוה לקנות ספרו, ולתמוך בו כדי שיוכל להמשיך ולהוציא לאור עוד ספרים חשובים, שנוגעים להלכה ולמעשה, ויבורך כל מי שיכניס את ספרו לביתו בחותם הברכות שבתורה.

[חתימה]

הסכמת הגאון הגדול
רבי משה קופשיץ שליט"א
רב בית הכנסת אברכים רוממה
ור"מ ישיבת קול יעקב

בס"ד ירושלים עיה"ק תובב"א,
כ"ב שבט התשנ"ה

הנה אתא לקמאי הרה"ג ר' אברהם עדס שליט"א בנן של קדושים גדולי תורה ויראה ואייתי מתניתא בידיה ספר "ויאמר אברהם" על הלכות בורר.

וכבר איתמחי גברא בחיבוריו היקרים על הלכות רבות בסדר מועד, וכעת מוסיף על הראשונים מפלפל ומסלסל בדברי הראשונים והאחרונים ומסיק שמעתתא אליבא דהלכתא.

והנה ראיתי כי דברים נכוחים למבין וישרים למוצאי דעת, ודבריו ישאו חן בעני לומדי התורה הלומדים בהלכות קשות אלו הלכות בורר.

בפרט שדבריו מסודרים סידור נאה דבר דבור על אפניו ובו טור פנימי אשר מקיפו החיצון וטעמו כיין ישן אשר דעת זקנים נוחה הימנו.

ועל זה באתי עה"ח

הסכמת הגאון מחשובי מורי ההוראה
פעיה"ק ירושלים ת"ו
רבי יחיאל מיכל שטרן שליט"א
מח"ס "כשרות ארבעת המינים"

בס"ד, כ"א תמוז תשנ"ד

חזיתי איש טהור, ואייתי ספר לידיה בשם ויאמר אברהם, לאותו אברך גדול בתורה ה"ה הרה"ג ר' אברהם חיים עדס, והוא מאותו נצר הדס שמשפחתו מבשמת ומבושמת, נכד לאותו גאון מופלא צדיק כתמר, ולבו היה נעים כאתרוג, וקירב אותן ערבות לכלל ישראל, ה"ה צדיק יסוד עולם רבי יעקב עדס זצוק"ל, ובנו של איש צנוע עושה ומעשה למען התורה, ות"ח גדול מיתקה, ה"ה הרה"ג רבי דניאל עדס שליט"א, ולא לחנם זכה המחבר להוציא ספר חשוב על עניני ארבעת המינים, שרבים יהנו לאורו, וכבר אמרו חז"ל כל שהוא ת"ח ובנו תלמיד חכם ובן בנו ת"ח לא יפסק התורה מפי זרעו, ויזכה להמשיך בהפצת תורתו ברבים.

הכו"ח למען כבוד התורה
יחיאל מיכל שטרן

Contents

LULAV

The Different Varieties of Lulav 2
Size Requirements 12
Separated Leaves 13
A Lulav whose Leaves have Fallen Off 14
A Lulav with a Split Middle *Teyomes* 18
A Lulav with Two Middle *Teyomos* 23
Ko're 25
A Dried-Out Lulav 29
A Lulav with a Cut-Off Top 31
Himnak 35
The Qualification of *Hadar* during Chol HaMoed 44
Hiddur Mitzvah — Beautifying the Mitzvah 44
A Wrinkled or Bent Lulav 45
A Lulav with a Hooked Tip 48
A Damaged Lulav 54
Attaching the Four Species 55

HADASIM

Triple-leaved *62*
The Minimum Length of the Hadas *68*
A Damaged Hadas *77*
Berries *83*
A Dried-Out or Withered Hadas *87*
Fragrance *90*
The Size of the Leaves *91*
A Hybrid Strain *92*
The Number of Hadasim Required *92*

ARAVOS

The Aravah *96*
The *Tzaftzafah* and the *Shitah* *101*
The Minimum Length of the Aravah *104*
A Dried-Out or Withered Aravah *105*
An Aravah with a Cut-Off Top *109*
Lavluv *113*
An Aravah whose Leaves have Fallen Off *116*

ESROG

Description of the Esrog *123*

Terms that Describe Discoloration of the Esrog *123*
Size Requirements of the Esrog *124*
A Hybrid Esrog *125*
Chazazis *132*
A Dried-Out Esrog *137*
A Punctured Esrog *139*
An Incomplete Esrog *144*
A Crushed Esrog *148*
A Cracked Esrog *148*
A Peeled Esrog *149*
The *Pitam* *151*
The Growth of the *Pitam* *159*
Pesticides *165*
A *Pitam* that Fell Off *165*
Unsightly Esrogim *170*
The Stem *174*
The Origins of Black Spots *175*
The Halachos of Black Spots Caused
    by Discoloration *179*
The Origins of White Spots *189*
The Halachos of White Spots Caused
    by Discoloration *191*
The Origins and Halachos of Brown Spots *192*
Additional Halachos of Brown Spots *196*
Halachos of Kosher and Pasul Spots *197*
The Secretion of the Esrog's Juices *199*
Insect Infestations *204*

Chotem	*205*
The Origins of *Bletelach*	*207*
The Halachos of *Bletelach*	*212*
A Pasul Spot that was Removed	*214*
Soaked or Cooked Esrogim	*214*
A Black or Orange Esrog	*215*
A Long and Thin Esrog	*217*
A Perfectly Round Esrog	*218*
An Esrog that was Grown in a Mold	*222*
Twin Esrogim	*223*
A Straight-Sided Esrog	*224*
A Green Esrog	*226*
Artificially Stimulating the Color Change of an Esrog	*229*
An Esrog that was Placed beneath a Bed	*234*

Preface

The feeling of excitement returns every year — the days following Rosh Hashanah. These days of inspiration, of *teshuvah* and *tefillah*, have begun to transform us and now we immerse ourselves in preparation for the *mitzvos* of Sukkos, which fills our time, our minds and hearts. The *arba'ah minim* begin to appear — a man walking with a lulav in the street or the Rav scrutinizing an esrog. We already anticipate holding them in our hands, that moment before reciting the *berachah* and *shehecheyanu*, before performing the *na'anuim* for the first time.

We start the search, with a tingle in our veins. A trip to the Hebrew bookstore, *arba'ah minim* market, or local supplier and there we stand before a large selection of the Four Species. We certainly have a general idea of what to look for or perhaps we have even learned through the halachos in depth. Even so, it can be difficult to apply that knowledge without understanding how the *arba'ah minim* grow and develop and without actually observing how the *poskim* decide what is *mehudar* and what is not.

How can we tell which varieties of lulavim are from the date palm? What kind of a black spot will make an esrog pasul? What about white spots? Are green esrogim kosher? What are the criteria for a hadas to be considered triple-leaved? Can one use a withered aravah? What exactly is a *himnak* or a *chazazis* or a *lavluv*? These and other practical questions must be answered by someone with hands-on experience, who has learned the *sugyos* thoroughly and acquired the knowledge that can only come from observing and attending *talmidei chachamim*.

It therefore is not surprising that the Hebrew sefer *Arba'as HaMinim LaMehadrin* became an instant best seller when it first appeared in Eretz Yisrael one year ago, praised and acclaimed by *Gedolei Yisrael*. The author, Rabbi Avraham Chaim Adess *shlita*, a renowned *talmid chacham*, presents to the reader the fruits of his many years of research under the tutelage and guidance of leading *poskim*. Every year before Sukkos, in the Bayit Vegan neighborhood of Yerushalayim, throngs of Yeshiva students and *avreichim* wait in line for hours to show Rav Adess their *arba'ah minim* and learn from him their status — whether they are *mehudar* or pasul, kosher *l'chat'chilah* or *b'dieved*. *Arba'as HaMinim LaMehadrin* presents this knowledge in written form, with over 170 color photographs to help illustrate the fine details of halacha. Through this work, readers can attain a deeper proficiency in the laws of the *arba'ah minim* and an enhanced observance of this special mitzvah.

For this reason, we felt it essential that this sefer be translated into English, to benefit the large English-speaking public. We are thankful to Hashem that we have completed this English translation in time for the upcoming Sukkos Festival. This sefer is certainly appropriate for those learned in the halachos of *arba'ah minim*; yet, one of our goals in adapting the sefer for a wider audience was that it should also become a source of halachic guidance for those who do not have direct access to experts in the laws of the Four Species and are incapable of determining the halachos on their own. With this aim in mind, many detailed footnotes have been omitted, along with the author's in-depth halachic analysis appropriate only for accomplished Torah scholars; for the most part, only those footnotes that relate to verbal communication from *Gedolei Yisrael* have been included.

We would like to express our gratitude to Rabbi Menachem Goldberger, who reviewed the entire translation for halachic accuracy, as well as to the entire editorial and production staff here at Feldheim Publishers in Yerushalayim. Special mention must be made of Mrs. Bracha Steinberg for her attractive design and Mrs. Tzviya Blass for her professional typesetting and page make-up. A big *yasher koach* to Rav Refoel Reichman for his constructive comments.

May the punctilious observance of this beautiful mitzvah bring increased *berachah* upon all of *Klal Yisrael*.

David Kahn
Editor

Lulav

"Take for yourselves on the first day … branches of a date palm tree."
(Vayikra 23:40)

The Different Varieties of Lulav

The Torah specifies that only a lulav from a tree similar to a date palm may be used. There are several kinds of palm trees in the date family that are certainly kosher. There are other kinds of palm trees that resemble the date palm in certain characteristics and therefore are of questionable kashrus.[1]

The Israeli lulav market handles only a small portion of the range of kosher lulavim: those grown in Jericho, El Arish, Beit Shean, and Jordan. Of these lulavim, the Jordanian lulavim are unique in their appearance. The other lulavim are thick, long, and green. Their leaves are either closely attached to the spine (*shidrah*) or slightly spread out; both forms are kosher. The leaves of the Jordanian lulav, however, always remain closely attached to the spine, giving the appearance of a thin, long, sparse leaved staff.

To perform the mitzvah of shaking the lulav (*na'anuim*) correctly, according to the opinion of the Ritva, one can use only a lulav whose leaves spread out. He requires that the leaves vibrate while the lulav is being shaken. This is very difficult with the tighter Jordanian lulav. Also, it is necessary to check the Jordanian lulav to ensure that it is at least possible to separate the leaves from the *shidrah*. If it is impossible to separate them, then they are considered part of the *shidrah*, and the lulav is considered leafless and totally unfit for use.

In recent years, a new variety of lulav known as *Deri* has appeared on the market. This lulav is renowned for its unique beauty and thickness, and it is kosher *l'chat'chilah*. Similarly, there are dozens of new kinds of palm trees that are kosher, al-

---

1. Chasam Sofer, *Sukkah* 32b, Chazon Ish, *Kilayim* 2:18.

though they were not known in previous generations.

On the other hand, there is a variety of palm imported from the Canary Islands known as the Canary palm, over which the opinions of modern *poskim* differ. Rav Shlomo Zalman Auerbach *zt"l* held that it is kosher, whereas Rav Moshe Feinstein *zt"l* held that it is pasul even *b'dieved*. Therefore, parents should instruct their children not to take lulav branches from the palms that are commonly planted in gardens and on the dividers of the highways in *Eretz Yisrael*, for very often they are Canary palms, and unless they are specifically known to be date palms, we should refrain from using them. (Obviously, it is forbidden to take a lulav branch from any tree without permission of its owner).

There are numerous differences between the date palm and the Canary, both in the appearance of the tree itself and in the appearance of the lulav branches. Regarding the tree itself, the date palm is tall and narrow, and its branches point upwards, whereas the Canary palm is shorter and thicker, and its branches point downwards. Furthermore, offshoots grow from the trunk of the date palm , whereas the Canary palm has no such branches.

Regarding the differences in their lulav branches, the leaves that stem from the date palm's lulav grow at a distance of 1.2-1.6 inches (3-4 centimeters) from each other, intermittently with smaller, 0.4 inch (1 centimeter) intervals, whereas the Canary palm's leaves all grow at a distance of one centimeter from each other. The leaves of the date palm end in a point, whereas the Canary palm has leaves that are not pointed. The middle leaf of the date palm lulav ends at a distance from the top of the spine, whereas the middle leaf of the Canary palm's lulav is shorter. When the date palm's lulav is held at an angle, it remains straight, since its spine is hard. However, the spine of the Canary

The date palm – note the tall, narrow trunk.

The Canary palm – the trunk is shorter and thicker than the date palm's.

Offshoots such as these grow from the trunk of the date palm.

The Canary palm does not grow offshoots.

The date palm – the spaces between the leaves of the date palm are wider than those on the Canary palm.

The Canary palm – the leaves of the Canary palm grow closer together.

The date palm – the middle leaf grows to a considerable distance from the tip of the shidrah (spine).

The canary palm – the middle leaf ends at a short distance from the shidrah.

palm's lulav is softer, and droops slightly when held at an angle. The leaves of the date palm's lulav are harder than the Canary palm's. Finally, the fruit of the date palm are elongated, whereas the fruit of the Canary palm are round.

We have only discussed the Israeli lulav market. The situation in other countries could very well differ, and requires independent research.

Size Requirements

The minimum length for a kosher lulav is four *tefachim*, which, according to the opinion of Rav Chaim Na'eh, measures 12.8 inches (32 centimeters), and according to the Chazon Ish 15.4 inches (38.4 centimeters). The *shidrah* of the lulav itself must be at least this length, not including the leaves that grow from its tip.

At the base of the lulav grow two leaves, one on either side of the lulav. The four *tefachim* must be measured from the higher of the two, since the side of the lulav across from the lower leaf is bare and the entire length of the four *tefachim* must be covered by leaves on both sides.[2]

---

2. The Mishnah Berurah speculates as to whether the entire lulav must be covered by leaves, or just the greater part of the lulav, and does not come to a clear decision (Bei'ur Halachah 645:1). Therefore, on the first day of Yom Tov when the mitzvah of taking a lulav is *mi-d'Oraysa*, a Torah commandment, we must be stringent and require the entire length of four *tefachim* to be covered by leaves. However, during the rest of Sukkos, when the mitzvah is *mi-d'Rabbanan*, Rabbinical origin, we may be lenient, and where no other lulav is available, a lulav whose greater part is covered by leaves will suffice. Therefore we may measure the four *tefachim* from the bottom-most leaf, although the lulav is not covered on both sides there, and regard as sufficient that the greater part of the lulav is covered on both sides.

Measuring the length of the lulav.

Separated Leaves

As a lulav ages, the leaves fan out and become hard. When it becomes too stiff to return to its position adjacent to the *shidrah*, the lulav is pasul. As long as it is possible to bend the leaves back to the *shidrah*, it is not necessary to actually do so, nor to tie them to the *shidrah* to keep them straight.

Some hold the opinion that it is preferable that the lulav's leaves lie flush against each other, with no space between them.

Rav Ezriel Auerbach interpreted this opinion to mean that even if the leaves spread out somewhat, if the leaves become pressed together when the lulav is shaken or lain flat on a table, it is still considered within the category of the preferable *mitzvah min ha-muvchar*.

Others hold that the lulav's leaves do not need to lie flush together to be categorized as *mitzvah min ha-muvchar*. However, if, when the lulav is held in an upright position, its leaves spread out and droop, but when held upside down, its leaves fall against each other around the *shidrah*, it is not considered a *mitzvah min ha-muvchar*, although it is still kosher.

A Lulav whose Leaves have Fallen Off

1. If the majority of leaves have fallen off a lulav, it is pasul. However, if only the minority of the leaves have fallen, and the lulav is still covered with leaves, it is kosher. A leaf is considered to have fallen whether it is completely detached or dangling and only slightly connected. Furthermore, a leaf that has fanned outward and hardened to the point where it is impossible to bend it back to the *shidrah* is also considered to be fallen.

2. A lulav from which the majority of leaves have fallen, yet the majority of the minimum necessary length is still covered with leaves (for example, a lulav of seven *tefachim*, four of which are left bare and three of which remain covered with leaves), is of questionable kashrus. If we were to look at the whole lulav as it is, we would see that the greater part is bare and it is pasul. However, if we were to take into account that the minimum lulav length need only be four *tefachim*, it could be argued that the three *tefachim* that are covered with leaves constitutes a majority of the length requirement, and it is kosher.

This uncertainty does not apply where the four bare *tefachim* are all at the bottom of the lulav. In such a case, the lulav would certainly be kosher, for then we could ignore the bottom three *tefachim* as if they were detached, leaving the remaining lulav mostly covered and therefore kosher.

3. A lulav with a broken *shidrah* is pasul, even if the two halves were reattached and bound together. However, if the two halves

A lulav whose shidrah has broken into two separate pieces is pasul.

remained slightly attached, it is permissible to bind them together to strengthen the connection, and such a lulav is kosher and a *berachah* may be recited over it.

A lulav whose shidrah has snapped yet the two pieces are still slightly connected.

Hashem wished to benefit the Jewish people, and provide us with every opportunity to perfect ourselves. We see that there are four distinct types of people among Klal Yisrael. There are those who have both the "fragrant smell" of mitzvos, and the "good taste" of Torah study. There are those who have neither. There are those who study Torah, but are not scrupulously observant of its mitzvos. There are those who keep the mitzvos, yet do not study Torah.

Hashem commanded us to take the Four Species: the esrog, which has both a good taste and a fragrant smell; the aravah, which has neither; the hadas, which smells good but has no taste; and the lulav, whose dates taste good but have no fragrant aroma. By binding them together, we show that each one completes the other's deficiencies. Just like the aravah, the Jew with neither Torah study nor mitzvah observance, also complements the group. Although such a Jew when viewed apart from the community might appear wicked, when united with the group he benefits them. His presence helps to sharpen the light of Torah and mitzvah observance.

We find this concept represented in the ketores, the incense offering that was brought in the Beis HaMikdash. It included galbanum, a substance that smells foul when burned alone, but when mixed with the other aromatic herbs, improves the fragrance. Therefore, our Sages warned us, "Any community prayer that does not include the prayers of the wicked, is incomplete."

(Seder Hayom)

A Lulav with a Split Middle *Teyomes*

The lulav grows in such a way that every pair of leaves is connected in the back. This gives the appearance of one leaf that is folded over in half. The point at which they meet, or what seems to be the fold, is called the *teyomes*, from the Hebrew root for "twin".[3]

1. A lulav whose middle *teyomes*, i.e. *teyomes* on its central leaf, is split down the greater part of its length is pasul on the first day of Yom Tov. For the rest of the days of Sukkos, however, it is kosher and a *berachah* may be recited over it.

2. Some are of the opinion that it is preferable to use a lulav whose middle *teyomes* is not split at all. Others hold that if the middle *teyomes* is split up to a *tefach*,[4] it

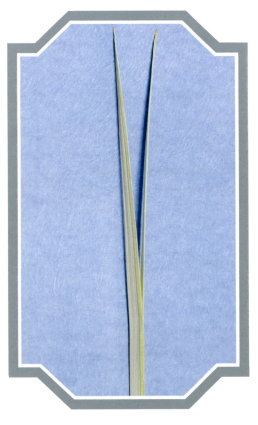

The back that connects the twin folds that make up each leaf is called the teyomes.

---

3. Rambam, *Hilchos Lulav* 8:4 .
4. 3.2 inches (8 centimeters) according to Rav Chaim Na'eh, or 3.8 inches (9.6 centimeters) according to the Chazon Ish.

is still considered a *mitzvah min ha-muvchar*. Some hold that even if the lulav fulfills the basic requirement, i.e. that the greater part of the middle *teyomes* is not split, it is still considered a *mitzvah min ha-muvchar*.

Therefore, the order of preference in purchasing a lulav should be as described above, that it is best to purchase a lulav whose central *teyomes* is not split at all. If such a lulav is unavailable, it is best to purchase one whose split is no more than a *tefach*. If this too is not possible, it is best to purchase a lulav whose central *teyomes* is not split for more than half of its length. This order of preferences is only for *mitzvah min ha-muvchar*; one would not have to borrow another person's lulav to fulfill the mitzvah of lulav according to a more stringent opinion. If, however, the only lulav available is one whose central *teyomes* is split for more than half of its length, it is pasul for the first day of Yom Tov and one must make a *berachah* over another person's lulav.

3. According to the opinions that the *mitzvah min ha-muvchar* would be to use a lulav whose middle *teyomes* is not split at all, this is only due to a concern that when the lulav is shaken the crack will continue to split. Therefore, the partially split tip of the *teyomes* (i.e. less than halfway down) may be mended with glue, and the lulav would then be considered *mitzvah min ha-muvchar* according to all opinions.[5]

4. For the days of Chol HaMoed, a lulav whose middle *teyomes* is partially split is still considered *mehudar*, while a lulav whose middle *teyomes* is split for the greater part of its length is still kosher, but not *mehudar*.

---

5. Heard from Rav Nissim Karelitz.

5. A child of under bar-mitzvah age may *l'chat'chilah* use a lulav whose central *teyomes* is split up to a *tefach*.

6. If a lulav has two middle *teyomos*, both of them must fulfill the qualifications outlined above. If even one of them is split down the greater part of its length, it is pasul according to the opinion of the Mishnah Berurah. The Chazon Ish (145:5) writes that there is room for leniency in regard to such a lulav.

7. A lulav with a single central leaf that is not paired to another leaf by a *teyomes* is pasul. Furthermore, even if the middle leaf is paired by a *teyomes*, if the majority of the other leaves are not paired, it is pasul as well.

8. If one leaf of the middle *teyomes* is shorter than the other, the lulav is still considered kosher and *mehudar*, provided that the difference between them is not obvious. (A difference of only two millimeters is certainly considered unnoticeable.[6]) If no other lulav is available,

One leaf of the teyomes is taller than its twin.

---

6. Heard from Rav Yosef Shalom Elyashiv.

then even if the difference is noticeable, the lulav is still kosher, provided that the shorter leaf covers at least half of the length of the taller one.

9. If the leaves of the middle *teyomes* are of equal length, but a thorn-like protrusion grows at an angle from the tip of one leaf, the leaves are still considered to be of equal length and the lulav remains *mehudar*. The thorn-like growth is not included in the length of the leaf.

The leaves of the teyomes are of equal height, but a thorn-like protrusion grows from one of them.

10. If the leaves of the middle *teyomes* are of unequal width, the lulav is still kosher, provided that the narrower leaf covers the greater part of the width of the wider leaf for the majority of its length.[7]

One leaf of the middle teyomes is wider than its twin.

---

7. Heard from Rav Shlomo Zalman Auerbach *zt"l*.

For the majority of the length of the teyomes, the narrower leaf covers the greater part of the wider leaf's width.

11. If the leaves of the lulav grow from only one side of the *shidrah*, leaving the other side bare, the lulav is pasul for the entire duration of Sukkos.

A Lulav with Two Middle *Teyomos*

Sometimes it is difficult to distinguish whether there is one middle *teyomes*, with all of the stringencies particular to the *teyomes* mentioned above, or two. One method of distinguishing is by the tiny holes that are found at the top of the middle *teyomes*. If there is only one middle *teyomes*, then each of its paired leaves should have a tiny hole at its tip. If there are two middle *teyomos*, then each *teyomes* should have only one hole that spreads across both of its leaves.

1. According to the opinion of the Geonim, a lulav with two middle *teyomos* that were separated from each other after the lulav was picked from the tree is pasul. If they were separated before the lulav was picked, it is kosher. However, since the Mishnah Berurah does not cite the opinion of the Geonim, it is unnecessary to abide by it, and the separation of the two middle *teyomos* does not invalidate the lulav.[8]

2. If even one of the two middle *teyomos* is split down the greater part of its

A lulav with two middle teyomos.

---

8. Heard from Rav Ezriel Auerbach.

length, it has the same status as a lulav with only one *teyomes* that is similarly split, and the lulav is pasul.

3. Below, in the section discussing a lulav with a cut off top, we will discuss the status of a lulav with two middle *teyomos*, one of which is cut off.

4. Below, in the section discussing *himnak*, we will discuss the status of a lulav with two middle *teyomos* that spread apart from each other.

Ko're (A Brittle, Brown Substance That Often Coats the Lulav)

1. Ashkenazim are careful not to use a lulav whose tip is covered with *ko're*. The *ko're* holds the leaves of the lulav together, and prevents them from shaking properly during the *na'anuim*, as it is performed among the Ashkenazim. Others, however, require only to wave the lulav during the *na'anuim*, and not to shake its leaves. Therefore the Sephardim, who follow this opinion, use a lulav that is covered with *ko're*.

However, it is advisable for those who wish to use a lulav that is covered with *ko're*, to choose specifically a lulav from the *Deri* palm, whose middle *teyomes* may be presumed to be unsplit, even without removing the *ko're* to check for certain. The other kinds of lulav, however, such as El Arish, Jericho, and Beit Shean, very often have split middle *teyomos*, and therefore it is better not to use them without first removing the *ko're* to ascertain the kashrus of the middle *teyomes*.

There are those who refrain from using even a *Deri* lulav that is covered with *ko're*. Although it is most likely that the middle *teyomes* is not split, they prefer to make certain that it does not have any of the other deficiencies mentioned above, such as one

A lulav whose tip is covered with ko're.

leaf being longer than the other, etc.

2. The Brisker Rav *zt"l* was careful not to use a lulav whose middle *teyomes* leaves were separated by *ko're* (even if the head of the lulav was not entirely enveloped in *ko're*). However, this stringency applies only if upon the removal of the *ko're*, the leaves of the middle *teyomes* appear forked, like a *himnak* (see below). If, after the *ko're* is removed, the leaves of the middle *teyomes* appear adjacent to one another, then it is kosher *l'chat'chilah* even according to the Brisker Rav.

3. Those who do not abide by the custom of the Brisker Rav, may use a lulav whose middle *teyomes* leaves would appear split like a *himnak* if not for the *ko're* between them.

4. As discussed below, a dried-out lulav is only pasul if a superficial glance clearly reveals its dryness. Sometimes, after the *ko're* is removed, a faint brown stain remains on the leaves of the lulav. If this stain masks the dryness of the lulav so that a superficial glance does not clearly detect it, the lulav is kosher.

5. If one wishes to remove the *ko're* from the lulav, it is important to be careful not to damage the lulav in the process. When choosing a lulav, one must be careful not to damage the seller's merchandise. A potential buyer who has damaged a lulav while inspecting it for purchase, must inform the merchant that he has done so and pay him for his loss.

A middle teyomes with ko're between its leaves.

A Dried-Out Lulav

1. A lulav is pasul if most of its leaves, for the greater part of their length, have dried out. Such a lulav is pasul even if the middle *teyomes* is still moist.

2. Even if most of the leaves are still fresh, some hold that if the tip of the middle *teyomes* is dried out to such an extent that a superficial glance clearly discerns its dryness, the lulav is pasul.

Some *poskim* hold a more lenient opinion and state that the dryness of the middle *teyomes* will invalidate the lulav only if it is dry along the greater part of its length. Some hold that the dryness of the middle *teyomes* will not invalidate the lulav even if the entire length of the leaf is dry.

It is best to purchase a lulav whose middle *teyomes* has no clearly visible signs of dryness at its tip. However, as long as the greater part of the middle *teyomes* is still moist, it is permissible to use the lulav, and a *berachah* may be recited.

3. The definition of dryness is that the green color of the lulav has entirely faded.

The tip of the teyomes has faded and lost its green pigment.

4. A dried-out lulav is pasul for the entire duration of Sukkos.

5. If the tips of the leaves have been burnt by the sun, as long as they are not so brittle as to crumble when rubbed by a fingernail, the lulav remains kosher. If the holes at the tip of the middle *teyomes* are still visible when viewed under a magnifying glass, this is a sign that the lulav is not dried out, and it is considered *mehudar*.[9]

The tips of the lulav's leaves have been burnt by the sun.

---

9. Heard from Rav Ezriel Auerbach.

6. If the tip of the middle *teyomes* is slightly dried, but it is difficult to discern the severity of the dryness, due to damage done by the sun, as long as a superficial glance does not detect clear signs of dryness, the lulav remains kosher.

A Lulav with a Cut-Off Top

1. It is better not to use a lulav whose middle *teyomes* is cut off at the top even slightly. If this is the only lulav available, it is permissible to use it, since R. Yosef Ka'ro (known as "the Mechaber" [i.e. author] of the Shulchan Aruch), and many other sources hold that it is kosher. However, a *berachah* should not be recited. If one did recite a *berachah*, it is not considered a *berachah l'vatalah* (a transgression of the prohibition against reciting invalid *berachos*), since there are those opinions that hold that such a lulav is kosher.

2. If the tips of the majority of the leaves that rise above the *shidrah* are cut off, the lulav is pasul, for the entire duration of Sukkos according to all opinions.

3. A lulav should be checked with a superficial glance for possible signs that the tip of the middle *teyomes* has been cut off. If the lulav appears to be intact, it is considered *mehudar*. If there is reason to suspect that the lulav's tip has been severed — for example, if specks are noticed at the top of the leaf that might have resulted from its being cut — the status of the leaf can be verified by inspection under a magnifying glass. If the small holes that are generally present at the tip of the middle *teyomes* are visible, it can be assumed that the tip has not been cut.[10]

---

10. Heard from Rav Nissim Karelitz and Rav Ezriel Auerbach.

4. The thorn-like protrusion that sometimes grows from the tip of the lulav is not considered to be part of the lulav. Therefore if it is cut off, the lulav remains *mehudar*. In order to ascertain that none of the lulav itself was cut, together with the protrusion, the tip of the lulav can be checked for the holes that would verify that the lulav remains intact and is kosher. If the holes are not visible, or if instead of a hole there seems to be a crack, then it is clear that the tip of the lulav was cut and the lulav should not be used.

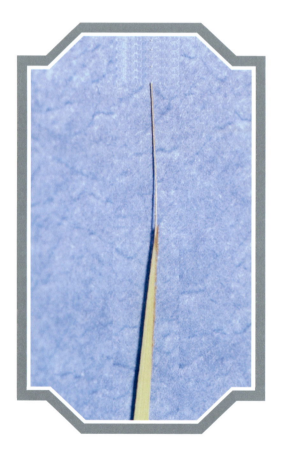

A lulav with a thorn-like protrusion at its tip.

5. If a lulav has two middle *teyomos*, one of which was cut off at its tip, *l'chat'chilah* the lulav should not be used, but *b'dieved* it is kosher.

6. Even if only one of the two leaves of the middle *teyomes* was cut off at its tip, the lulav should not be used.

7. If the middle *teyomes* is broken but not entirely cut off, even if the tip of the lulav droops downward due to the fracture, the lulav remains kosher. However, in such a case a reasonable concern exists that the tip of the lulav was cut off as well, and it must be checked by inspecting the holes at its tip, as discussed above.

8. Some opinions invalidate a lulav with three middle *teyomos* (i.e. all three extend from the tip of the *shidrah* and not from its sides) if any two of them have been cut off at the top. Others hold that the only *teyomes* taken into consideration is the middle of the three.

A lulav with three middle teyomos.

Himnak

1. If the leaves of the middle *teyomes* diverge from each other to the extent that they are noticeably separate, the lulav is pasul even without the conditions discussed above of being split down a *tefach*'s length of the *teyomes*. This condition is referred to as a *himnak* split.

Himnak – the leaves of the middle teyomes diverge from each other to the extent that they are noticeably separate.

2. Some hold that a *himnak* split will invalidate the lulav, regardless of whether it is a result of natural growth or of damage that was later incurred.

Others hold that if the *himnak* split is a result of damage, and the two leaves of the *teyomes* rejoin when the lulav is shaken during *na'anuim*, or placed on a table, the lulav is kosher. Furthermore, if the two leaves of the *teyomes* can be easily pressed together, even if they immediately reopen when the pressure is removed, the lulav remains kosher.

3. The Chazon Ish held that a lulav is considered to have a *himnak* split only when its leaves grow in opposite directions. If, however, they remain parallel, even if separated by a distance of up to one millimeter, the lulav remains kosher.

Although the leaves of the middle teyomes are separate, they run parallel to one another.

Although the leaves of the middle teyomes are separate, they run parallel to one another.

The left leaf of the middle teyomes diverges to the side. This lulav has a himnak split and is pasul.

The right leaf of the middle teyomes diverges to the side. This lulav has a himnak split and is pasul.

The leaves of the middle teyomes diverge from one another. Such a lulav is pasul.

4. A *himnak* split will invalidate a lulav only when it is obviously visible at a superficial glance. If, however, the *himnak* split is apparent only under close scrutiny, the lulav remains kosher. Whatever can be seen when the lulav is held at a distance of 12 inches (30 centimeters) is considered obviously visible.

5. A lulav with two middle *teyomos* that grow apart from each other, yet the leaves of each one individually are closed, is kosher. Similarly, if the lulav has only one middle *teyomes* whose leaves are closed, yet it grows way from its adjacent leaf so that it appears to have a *himnak* split, it is also kosher. The lulav is considered to have a *himnak* split only when the twin leaves of the middle *teyomes* grow apart from each other.

6. A lulav whose middle *teyomes* is shorter than the adjacent *teyomos* is kosher, and is in fact preferable as a *mitzvah min hamuvchar*. Since it is shorter, the neighboring leaves guard it and preserve its freshness.[11]

Why did Hashem command us to use these Four Species specifically? The esrog resembles the heart, and atones for the sinful thoughts of the heart. The hadas resembles the eyes, and atones our searching out for sin, as the verse says, "And you must not be searching, following... your eyes after which you [tend to] stray." (Bemidbar 15: 39). The aravah resembles the lips, and atones for inappropriate speech. The single seed of the palm tree's date represents the singular desire of our heart to serve our Father in Heaven.

(Orchos Chaim, R'Aharon HaKohen of Lunil)

---

11. Heard from Rav Chaim Kanievsky.

The middle teyomes is shorter than the adjacent leaves on both sides. Such a lulav is kosher.

The Qualification of *Hadar* during Chol HaMoed

Many of the requirements for a kosher lulav derive from the need for the lulav to be *"hadar"*. Although the verse in *Vayikra* (23:40) mentions this in regard to the esrog (*"pri eitz hadar"*), it actually applies to all of the Four Species as well.

As a general rule, we can state that, with the exception of damage (i.e. incompleteness), the lulav with a split middle *teyomes* or *himnak* split, and the edibility of the esrog, all other factors that determine whether one of the Four Species is kosher or not are the result of its need to be *hadar*.

1. The Mechaber holds that the qualification of *hadar* applies only to Yom Tov,[12] whereas the Rema holds that it applies for the entire duration of Sukkos. The Ashkenazim customarily follow the rulings of the Rema, whereas the Sephardim follow the Mechaber.

However, since a lulav that lacks the qualification of *hadar* is unfit for the recitation of a *berachah*, the general rule of *safek berachos lehakel* (in cases of halachic uncertainty, it is forbidden to recite a *berachah*) applies. Therefore, the Sephardim must also refrain from using such a lulav at all, for, otherwise, they will be forfeiting the ability to recite the *berachah* upon taking them. [13]

Hiddur Mitzvah — Beautifying the Mitzvah

We should not confuse the above concept of *hadar*, with the universal requirement of *hiddur*. Our Sages (*Shabbos* 133b)

---

12. If it is difficult to find a different lulav, it may be taken on the second day of Yom Tov, but without a *berachah* (Shulchan Aruch 649:5).
13. Kaf HaChaim 649:57

interpret the word *"ve'anveihu"* (*Shemos* 15:2) to mean that we are obligated to perform mitzvos in a beautiful way. One of the examples given there is to take "a beautiful lulav." This requirement is codified in the Shulchan Aruch (Orach Chaim 656) and applies to all mitzvos, including each one of the Four Species. This means that, although a particular lulav or esrog may be technically kosher, we should spend up to one third of the price more for a nicer specimen so that it satisfies the requirement of *hiddur mitzvah*.

A Wrinkled or Bent Lulav

1. A lulav whose *shidrah* is bent forward or to the side is pasul, since this is not the natural way that a lulav grows. However, if the *shidrah* bends backward it is kosher, since this is the natural way that it grows. The side of the lulav where the *shidrah* is visible is considered its back.

According to some opinions, a lulav is considered bent and pasul only when it arches to such an extent that it is a bow-like half circle. Other opinions are more lenient and hold that it is pasul only if it bends into a half-circle and then contorts again to restraighten, concluding in its original straight line.

However, it is preferable to obtain a lulav that is not bent at all, for the sake of *hiddur mitzvah*. If the bend is only visible upon close inspection it is negligible, and the lulav is considered beautiful and *mehudar*.

2. If the tip of the middle *teyomes* ends with a wave-like zigzag, it is considered a perfectly normal, kosher leaf.[14]

---

14. Heard from Rav Chaim Kanievsky.

A slightly bent lulav is kosher, but lacks hiddur mitzvah.

3. If the middle *teyomes* divides into two separate pairs of leaves that branch out in a V-like shape, they are considered as two separate middle *teyomos* and the separation between them does

not invalidate the lulav as long as the leaves of each individual *teyomes* are closed. Some hold that such a lulav is considered wrinkled and therefore pasul. However, the accepted ruling is that it is no worse than the aforementioned wave-like zigzag, and therefore such a lulav is kosher.

The middle teyomes ends with a wave-like zigzag. Such a lulav is kosher.

The middle teyomes divides into two separate pairs of leaves that branch out in a V-like shape.

A Lulav with a Hooked Tip

1. If the tip of the middle *teyomes* of the lulav bends downward in a hook-like shape, in what is referred to colloquially as a *kenepel* (Yiddish for "button"), according to some opinions this is *mehudar*, for the "hook" protects the leaves and ensures that they do not separate. Others hold that such a lulav is pasul. Even if *ko're* covers the tip of the lulav such that the hook's tip is concealed, these *poskim* maintain that such a lulav is still pasul. If the tip is not really hooked but it is only slightly curved, all opinions agree that the lulav is kosher.

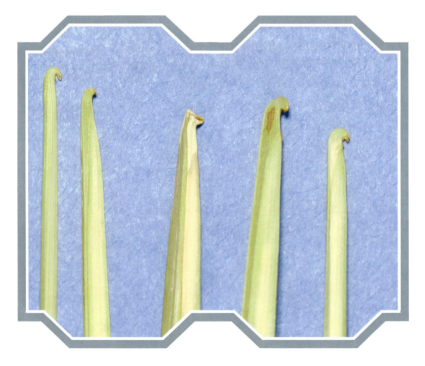

The tip of the middle teyomes of the lulav bends downward in a hook-like shape in what is colloquially called a " kenepel" (a button).

It is preferable not to use a lulav with a hooked tip, but if no other lulav is available the accepted ruling is that it is kosher. If, however, the middle *teyomes* is bent over in the middle of its length and not just at the tip, such that its leaves double over in two, it is pasul according to all opinions.

2. Similarly, if the majority of the tips of leaves curve into hook-like shapes, there is a difference of opinion as to the kashrus of the lulav. Here too, it is preferable not to use such a lulav, but if no other lulav is available, it may be used.

3. If the middle *teyomes* does not hook, but the adjacent one does hook, according to all opinions the lulav is kosher and *mehudar*.

4. According to the opinion that a hooked middle *teyomes* is *mehudar*, if a sizable minority of the other leaves hook as well, some maintain that this is also *mehudar*, while others hold that it is kosher but not *mehudar*.

6. If the middle *teyomes* does not curve into a full hook, but rather curves into a perpendicular point, like the Hebrew letter *vav*, the lulav is kosher according to all opinions.[15] The same is true if the middle *teyomes* has only a minute hook.[16]

---

15. Heard from Rav Yehuda Adess and Rav Nissim Karelitz. Rav Ezriel Auerbach said that such a lulav is *mehudar*.
16. Heard from Rav Nissim Karelitz.

The middle teyomes has a minute hook and is kosher according to all opinions.

The middle teyomes has a minute hook and is kosher according to all opinions.

The rightmost leaf is bent into a "vav" shape and is kosher according to all opinions. The middle and the leftmost leaves are beginning to bend into a full hook. There is a difference of opinion as to their kashrus.

7. A lulav whose leaves are deformed and strange-looking is pasul even according to the aforementioned lenient opinion that a lulav with a hooked tip is *mehudar*. [17]

---

17. Heard from Rav Nissim Karelitz.

A lulav whose leaves are deformed.

A Damaged Lulav

If a portion of the lulav's *shidrah* is damaged and missing, as long as the greater part remains, it is kosher.

A lulav that has been damaged and a portion of the shidrah is missing.

Attaching the Four Species

1. Some follow the custom of attaching the hadasim to the lulav's right side, and the aravos to the lulav's left, using a small holder woven from the leaves of a lulav.

According to the custom of the Ari, one hadas and one aravah are placed to the lulav's right, one hadas and one aravah to its left, and one hadas on the lulav's *shidrah*, with its tip leaning slightly towards the right.

A holder woven from the leaves of a lulav.

The Four Species bound together according to the custom of the Rema.

The Four Species arranged according to the custom of the Ari.

2. The Four Species should be arranged in such a way that the *shidrah* of the lulav rises at least one *tefach*[18] above the hadasim and the aravos. Furthermore, the hadasim should rise slightly higher than the aravos. Also, the entire length of the minimum three kosher *tefachim* of the hadasim and aravos should be placed above the bottom of the lulav.

The end of the shidrah

The top of the spine of the lulav is one tefach higher than the tops of the hadasim.

---

18. 3.2 inches (8 centimeters) according to Rav Chaim Na'eh, or 3.8 inches (9.6 centimeters) according to the Chazon Ish.

The Four Species are bound with the tops of the hadasim higher than the tops of the aravos.

3. The lulav, hadasim, and aravos should be bound together before Yom Tov with a double knot. It is forbidden to tie such a knot on Yom Tov. Therefore, if one forgot to do so before Yom Tov, according to the opinion of the Mechaber a simple bow should be used, in such a way that is permissible on Yom Tov. According to the Rema, a lulav leaf should be wrapped around them, and the ends of the leaf then tucked into its circle.

> *We awaken a person by shaking him back and forth, and the shaking of the lulav is meant to awaken those whose souls are in a spiritual slumber. On Shabbos, however, it is unnecessary to awaken the soul through the use of the lulav, for the holiness of Shabbos itself awakens us, as we say in Kabbalas Shabbos, "Awaken! Awaken! For your light has arrived."*
>
> *(Beis Avraham)*

Hadasim

Our Sages taught that the passage "branch of a braided [leaved] tree," refers to the hadas, whose leaves overlap, covering the wood of the branch. [Succah 32a]

Triple-leaved

The Torah does not prescribe the hadas by name, but rather refers to it as "branch of a braided [leaved] tree" (*Vayikra* 23:40). Our Sages received a tradition from Sinai that this refers to the hadas, whose leaves overlap, covering the entire length of the branch.

1. The hadas referred to in the Torah is triple-leaved, that is, the leaves grow in groups of three or more, emerging from adjacent

The leaves of the hadas grow from adjacent bases on its branch.

bases on the branch. Each group of three leaves is referred to as a "*ken*" or nest. The bases of the leaves of each *ken* must be level with one another in order for the hadas to be kosher. (We use here the term "base" to describe the connecting point of a leaf to its branch, i.e. the place on the branch from where the leaf emerges.)

2. If the group of bases of a *ken* are not perfectly level, but are close enough such that some part of each base overlap the others, the hadas is considered triple-leaved and kosher.

The bases are not perfectly level, but a line of circumference can be drawn connecting the roots of all three stems.

3. In a case of doubt as to whether the bases are level, the Chazon Ish was lenient, and the Brisker Rav was stringent.

It is unclear whether the roots of these stems are level.

4. If two leaves emerge from adjacent bases, and the third leaf emerges from a slightly higher or lower base, such that it is not level with the other two, the hadas is termed a *hadas shoteh* ("a foolish hadas"), and is pasul for the entire duration of Sukkos.

A hadas shoteh on which two leaves protrude from adjacent bases, but the third leaf protrudes from a slightly lower base.

5. If the leaves grow in groups of two, rather than three, the hadas is also considered a *hadas shoteh*, and is pasul for the entire duration of Sukkos.

A hadas whose leaves grow in groups of two.

6. If each *ken* is made up of seven leaves, and four have fallen off, the hadas is still kosher. If five fall off, the hadas is pasul.

7. If the leaves of the *ken* emerge from bases at equal heights on the branch, but one of the leaves is cut widthwise, making it shorter than the other two, the hadas is still kosher. The only stringency in the proximity of the leaves to each other is in regard to the leaves' stems, i.e. that the bases of each triple-leaf overlap.[1]

All three leaves of the ken grow from adjacent bases yet one has been cut across its width, making it shorter than the others.

---

1. Heard from Rav Nissim Karelitz.

The Minimum Length of the Hadas

1. A hadas must be at least three *tefachim* in length.² The preferable *mitzvah min ha-muvchar* is to use a hadas that is triple-leaved for the entirety of its length, according to the guidelines described above.

Measuring the length of the hadas.

---

2. 11.5 inches (28.8 centimeters) according to the Chazon Ish, and 9.6 inches (24 centimeters) according to Rav Chaim Na'eh.

2. The three *tefachim* are measured from the last triple-leaved *ken* (which excludes the branch directly below it), until the end of the branch (and not until the tip of the highest leaves). If the tip of the branch is green, as opposed to the typical woody brown, there is a difference of opinion among the *poskim* as to whether or not the green part is included in the minimum length.

The tip of the branch is green.

3. Even if the three triple-leaved *tefachim* of the hadas are not consecutive, they may be combined to render the hadas kosher *l'chat'chilah*.

The leaves have fallen off of the top and the middle, yet there remain three tefachim of non-consecutive leaves.

The esrog resembles the heart, the resting place of man's thought, which teaches us that we must serve Hashem with our mind.

The lulav resembles the spine, upon which the body is built. This teaches us that we must straighten all the actions of our body to properly serve Hashem.

The hadas resembles the eyes, which teaches us that we must not stray after our eyes.

The aravah resembles the lips, with which we finalize all of our decisions.

(Sefer HaChinnuch)

4. Even if the minimum length of three *tefachim* is not entirely triple-leaved, but the greater part of it is triple-leaved, although no longer *mitzvah min ha-muvchar*, the hadas remains kosher *l'chat'chilah*. In this case there is a difference of opinion among the *poskim* as to whether or not the "greater part" is calculated according to the length of the hadas,[3] or according to the number of nests.

Since this controversy remains unresolved, on the first day of Yom Tov, we must be stringent and ascertain that the greater part is triple-leaved according to both estimates. However, for the rest of the days of Sukkos, when the *mitzvah* is *mi-d'Rabbanan*, we may be lenient. Thus if the majority of the nests are triple-leaved, the hadas is kosher, even if the greater part of its length is not.

5. If a hadas measures four-and-a-half *tefachim* in length, two of which are triple-leaved, there is a difference of opinion among the *poskim* as to whether or not it is kosher. Some consider the two triple-leaved *tefachim* as the majority of the minimum three *tefachim*, and therefore the hadas is kosher. Others consider the two triple-leaved *tefachim* as the lesser part of the actual length of the hadas, and therefore the hadas is pasul.

---

3. The appropriate method of measuring the length of the hadas to determine which portion of it is triple-leaved is to locate every triple-leaved nest, and to measure from there to the next nest above it, even if that second nest itself is not triple-leaved.

These triple-leaved nests must constitute the greater part of the three *tefachim* branch within which they grow. Here, the three *tefachim* of the branch should be measured not from the bottom nest, but from the branch directly below it. This is in contrast to the method described above in paragraph 2. (Heard from Rav Refoel Reichman)

6. Even if the triple-leaved majority is not consecutive, the hadas remains kosher and a *berachah* may be recited.

7. Although we are lenient in allowing a mere majority of the three *tefachim* to be triple-leaved, the hadas itself must be at least three *tefachim* in length.

8. It is not necessary for it to be openly apparent that the majority of the leaves are triple-leaved, as long as a close inspection reveals that even a slight majority are triple-leaved.

9. It is not necessary for the uppermost *ken* to be triple-leaved. Even if the uppermost leaves have fallen off entirely, the hadas remains kosher.

10. If the leaves of each *ken* are too short to reach the following *ken*, yet the majority of the woody branch of the hadas is covered by leaves, some hold that the hadas is kosher, while others hold that it is pasul.

11. Even if the leaves do not remain tightly closed around the branch, yet spread outwards revealing the woody part of the branch, the hadas is kosher. Some, however, follow the custom of using only a hadas whose leaves remain tightly closed. The Chazon Ish held that as long as the leaves are soft and flexible enough to be bent back to the branch, it is as if the leaves were tightly closed.

12. If the hadas was originally triple-leaved, and later one leaf from each *ken* fell off, the hadas is kosher *b'dieved*. A *berachah* may be recited over it, if no other hadas is available.

A hadas without leaves at its top.

The tops of the leaves of one ken do not reach the stems of the ken above it.

The leaves of the right-hand hadas are tightly closed around the branch, whereas the leaves of the left-hand hadas spread outwards.

A Damaged Hadas

1. A hadas whose top leaves have been cut widthwise, leaving the entire base of the leaf intact, is kosher, provided that the woody branch of the hadas has not been damaged. However, it is a preferable *mitzvah min ha-muvchar* to use a hadas whose top leaves are not damaged in this way.

2. If the top of the woody branch of the hadas is cut, and the damage is clearly visible at a superficial glance, it is pasul according to all opinions. If the leaves of the hadas cover the place of the damage, then the damage is not considered clearly visible.

A hadas whose top leaves have been cut widthwise.

Therefore, the hadas is kosher and a *berachah* may be recited. However, it is preferable to use an undamaged hadas, if it is available.

The top of the branch has been cut and the damage is clearly visible.

The top of the branch has been cut but the damage is hidden by the leaves.

3. Although, generally speaking, a hadas is kosher regardless of whether the leaves remain tightly closed around the branch, or whether they stretch outwards, in the case mentioned above, in which it is necessary that the leaves hide the damage, the leaves must close tightly around the branch.

4. We have mentioned that if the top of the woody branch of the hadas has been cut and the damage is clearly visible at a superficial glance, the hadas is pasul. In such a case, it is permissible to cut off the damaged portion of the hadas in such a way that the leaves cover the new cut. In this way, the hadas is rendered kosher, on condition that after the damaged portion has been cut off, there remains the minimum three *tefachim* length described above.

5. If the top of the woody branch of the hadas is cut, and an addition to the branch then grew to cover the place of the damage, the hadas is kosher.

The top of the branch was cut and an addition to the branch then grew to cover the place of the damage.

6. Small branches growing from the hadas that are adjacent to its leaves should be removed.

A small branch growing between the leaves of the hadas.

Many years ago, during the time of the Vilna Gaon, the entire region of Lithuania suffered from a shortage of mehudar hadasim for Sukkos.

One day, a student of the Vilna Gaon passed by a house and noticed a bunch of beautiful hadasim by the window. He knocked on the door, and a gentile woman answered. He told her that he had seen the hadasim through the window and wished to purchase them. The gentile woman refused to sell them, explaining that the date for her daughter's wedding was approaching and the hadasim had been especially prepared for her daughter's wedding attire.

The Gaon's student explained his pressing need to acquire the hadasim in order to enable his rebbe, the Vilna Gaon, to perform the special mitzvah of arba'ah minim. At that time, the Gaon was renowned in Vilna among Jews and Gentiles alike as a venerable holy man. The woman had also heard of the Gaon's reputation. When she heard that the young man needed the hadasim for the Vilna Gaon, she gladly offered them for free, on condition that she would receive the merit of the Gaon's mitzvah of arba'ah minim. The student pleaded that she take money instead, raising his offer again and again, but she remained adamant on her condition.

The student realized that Sukkos had almost arrived, so he went to tell the Gaon. When the student explained to him the terms of the purchase, the Gaon joyously accepted them. "I have no desire for the reward of the mitzvah," he explained. "My only desire is to fulfill the will of the Creator."

The residents of Vilna testified that they had never seen the Gaon shake his lulav bundle more joyously than he did that year. "At last," he explained, "I can perform Hashem's mitzvos with no ulterior motive of receiving rewards."

(HaGaon HeChasid Mi-Vilna)

> *The Midrash compares the hadas to eyes and the aravos to lips. Just as the eyes are higher than the lips, it is proper to bind the hadasim to the lulav bundle higher than the aravos.*
>
> *(Bei'ur HaGra 651:1)*

Berries

1. If the hadas has green berries, it is kosher even if the berries outnumber the leaves. If, however, the hadas has black or red berries, it is kosher only if the leaves outnumber the berries. The calculation must be made strictly according to the number of leaves and berries, and not according to their size.

A hadas with green berries.

A hadas with red berries.

2. It is permissible *l'chat'chilah* to pick the black or red berries from the hadas, even after the hadas has been bound to the lulav and the aravos, provided that this is done before Yom Tov. On Yom Tov or Shabbos, it is forbidden to remove the berries, because this is considered "creating a utensil," a violation of the laws of Shabbos and Yom Tov. If a person transgressed the laws of Shabbos and Yom Tov and removed the berries, the hadas nevertheless becomes kosher. It is permitted to remove the berries on Chol HaMoed.

3. If a person forgot to remove the black or red berries before Yom Tov, and on Yom Tov wishes to use the hadas, he may ask a friend to pick the berries off and eat them. Since the friend's intention in this case is not to make the hadas kosher, but rather to eat the berries, it is not considered a violation of the prohibition against making a utensil. However, the owner of the hadas himself may not do so. *His* intention is obviously to make the hadas kosher, and therefore the act of picking the berries just to eat them is a ruse and unacceptable.

If the owner has other kosher hadasim that he may use instead, then he may pick the berries off and eat them himself, since his action is not done for the sake of preparing the hadas for use.

4. If a person forgot to remove the black or red berries before Yom Tov, and the above-mentioned suggestion is impractical, and another hadas is unavailable, he should nevertheless use the hadas. Although the hadas is pasul, one should still use it (without reciting a *berachah*), in order that the *mitzvah* of taking the lulav bundle not be forgotten.

The following incident illustrates to us the attitude of Gedolei Yisrael in performing hiddur mitzvos:

Once, a relative of the Steipler Gaon zt"l brought him six choice hadasim, telling him to choose from among them the three nicest ones. The Steipler realized that the relative intended to use the remaining three for himself, and insisted that the relative choose first. Although he had in fact intended to use them himself, he surreptitiously hinted that he had other hadasim at home, in order that the Steipler might not hesitate in choosing the best for himself. With great joy, and love of mitzvos, he chose the three nicest hadasim and placed the rest on the side. Almost immediately, the Steipler realized that his relative did not have other hadasim at home, and again insisted that the man take the nicest hadasim for himself. The relative demurred, explaining that even the three remaining hadasim were far more beautiful than the hadasim he was accustomed to using. Even so, the Steipler insisted that since the relative had troubled himself to acquire the hadasim, he must take the best for himself. The Steipler assured him that his Yom Tov could only be truly joyous if he knew he was not taking advantage of another person. The Steipler was adamant in his demand, and the relative was forced to take the nicer hadasim for himself. The Steipler's face then lit up, having succeeded in persuading his relative, and thanked him profusely.

This was the Steipler's custom every year in regard to arba'ah minim and all other mitzvos as well. He would never take the best of anything for himself and let the person who brought him things suffice with less.

A Dried-Out or Withered Hadas

1. A dried-out hadas is pasul for the entire duration of Sukkos. Two conditions render a hadas dry and therefore pasul. One is that the hadas has lost its color, and has faded into a white or brown. The second is that it crumbles when the pressure of a fingernail is applied. If only one of these two conditions exists, the hadas remains kosher.

A green hadas that has dried to the extent that it crumbles when the pressure of a fingernail is applied to it.

A hadas that has lost its green color and also crumbles under the pressure of a fingernail.

2. A withered hadas is kosher, as long as its leaves do not droop downwards when the branch is held upright.

A withered hadas.

3. If the entire hadas is dry except for the top *ken*, which remains moist, it is kosher. If the entire hadas is completely dry, and the top *ken* is only withered, *l'chat'chilah* it should not be used, but if no other hadas is available it is kosher.

The hadas is dry along the entirety of its length except for the top ken, which remains moist.

4. If the entire hadas is moist, except for the top *ken*, which is dry, it should not be used *l'chat'chilah*, but if no other hadas is available it is kosher. It is permissible to remove the top, dry *ken*, thus rendering the hadas kosher *l'chat'chilah*.

Fragrance

A hadas that does not have a pleasant fragrance is of questionable kashrus.

The Size of the Leaves

Some follow the custom of not using a hadas whose leaves are as large as two average thumbnails. This stringency applies only if all the hadasim that grow from a certain bush have this property. If, however, hadasim with small leaves grow on the bush as well, then those with larger leaves are also kosher.

Hadas leaves the size of two thumbnails.

A Hybrid Strain

1. Some *poskim* permit the use of a hadas that has been grafted together with a different species, as long as the resulting strain is similar to that of a non-hybrid hadas. The Mishna Berurah, however, is among those who forbid the use of any hybrid strain.

The Number of Hadasim Required

1. *L'chat'chilah* it is best to use three fresh hadasim. If neces-

A minimum of three hadasim must be used.

sary, it is sufficient to use three dry hadasim, if the top *ken* of at least one of them is still moist.

2. Some hold that it is permissible to add additional hadasim and aravos to the minimum required number. Of these, some hold that only triple-leaved hadasim may be added. The custom of Yemenite Jewish communities is to add even *hadasim shotim* to the three required hadasim.

The Yemenite custom is to add additional hadasim.

3. The ruling of the Shulchan Aruch is not to exceed the minimum prescribed number of three hadasim and two aravos.

Aravos

Our Sages taught that the aravah plant is referred to as "brook willows" in the verse since it usually grows by the bank of a brook. Another explanation is that its leaves are long and thin like a brook. However, Chazal derive from the same verse that aravos that grow in fields or on mountains are kosher as well.
[Succah 33b]

The Aravah

1. A kosher aravah has three signs:
   a. Its branch is red (see paragraph 3).
   b. Its leaves are thin and long.
   c. The edges of the leaves are smooth, and not serrated.

The three signs of the aravah:
a. the red branch   b. the thin, long leaves
c. the smooth edges of the leaves

2. Although the verse in *Vayikra* (23:40) specifies "brook willows", and indeed the aravah plant is generally found growing next to streams, it is not necessary for the kashrus of the aravah that it be found there.

3. If the branch of the aravah is still green, it is kosher. Although we stipulated above that the branch must be red, the Mishnah Berurah explains that when a green branch is exposed to sufficient sunlight it turns red. A white branch, however, is pasul.

An aravah with a green branch.

4. If the edge of the leaf is only slightly serrated, it is kosher.

Aravah leaves with slight serrations at their edges.

Smooth aravah leaves with no serrations at their edges.

The accepted custom that we have received from our forefathers is as follows: Upon the conclusion of Sukkos, after the Hoshana Rabbah davening, each person carries home his aravos and places them at the head of his bed to show his great love of the mitzvah. This is the appropriate practice.

(Tanya Rabbasi, 86)

The Rema writes that the custom is to keep the aravos from Hoshana Rabbah until Erev Pesach, and then use them as firewood for baking matzos. Thus, what was once used for a mitzvah will again be used for a mitzvah.

(Shulchan Aruch 664:9)

The Rivak followed the custom of reusing the aravos from Sukkos as quills for writing a sefer Torah, or as firewood for burning the chametz on Erev Pesach.

(Hagahos Maimoneos)

The aravos used on Hoshana Rabbah are a segulah for protection during one's travels. Their effectiveness depends upon the righteousness of one's deeds and the purity of his intentions. "He who trusts in Hashem, will rest in His shade."

(Menoros HaMaor, third candle, 4:6:7)

The *Tzaftzafah* and the *Shitah*

1. The *tzaftzafah* (poplar) is a tree whose branches slightly resemble the aravah, but are nevertheless pasul. The three signs of the *tzaftzafah* are as follows:
   a. Its branch is not red.
   b. Its leaves are round.
   c. The edges of its leaves are deeply serrated, like a saw.

2. The *shitah* (acacia) commonly grows along the highways in *Eretz Yisrael*. Its branches are pasul, even though their leaves also resemble the leaves of the kosher aravah.

The shitah plant that commonly grows along the highways in Eretz Yisrael.

The branch of the shitah plant is similar to the aravah, yet is pasul.

The Gemara differentiates between the kosher aravah and the tzaftzafah (poplar), which is posul. One of the characteristics that distinguishes between them is that the leaves of the aravah are smooth around the edges, while the leaves of the poplar are serrated.

Our Sages tell us that the aravah has neither a good taste nor a fragrant aroma. It represents the Jew who is neither learned nor dutiful in his mitzvah performance. By binding the aravah together with the other species, which represent the more righteous Jews, the aravah is uplifted and perfected. However, this is only possible for an aravah without serrated edges. So too, an unlearned Jew who respects Torah scholars, can bind himself to them and thus be uplifted. However, an unlearned Jew who mocks Torah scholars with sharp, derogatory comments, cannot be bound to them.

(Chiddushei HaRim)

The Minimum Length of the Aravah

1. The minimum length for a kosher aravah is three *tefachim*.[1]

2. The three *tefachim* are measured from the bottom of the branch[2] until its top, not including the leaves that protrude from the top.

Measuring the length of the aravah.

---

1. 11.5 inches (28.8 centimeters) according to the Chazon Ish, and 9.6 inches (24 centimeters) according to Rav Chaim Na'eh.

2. Although in regard to the hadas we stipulated that the three *tefachim* are measured from the last triple-leaved *ken*, here they may be measured from the

A Dried-Out or Withered Aravah

1. A dried-out aravah is pasul. The definition of dryness in this context is that the green color of the aravah has faded into a white or brown, and that it is so brittle that it crumbles under the pressure of a fingernail. If only one of these two deficiencies exists, the aravah remains kosher.

A green aravah that has dried to the extent that it crumbles when the pressure of a fingernail is applied to it.

---

very bottom of the branch. The reason for this difference is that it is essential for the kashrus of the hadas that it be entirely covered with triple-leaved nests, as we explained above; therefore we must measure from the bottommost *ken*, and not from the bottom of the branch. (Heard from Rav Nissim Karelitz)

106 THE FOUR MINIM

An aravah that has lost its green color and also crumbles under the pressure of a fingernail.

2. If the majority of the leaves are dry, the aravah is pasul for the entire duration of Sukkos, according to all opinions.

3. If the leaves of the aravah are slightly withered, the aravah remains kosher. If the head of the branch droops downwards, the aravah is likewise kosher. However, if most of the leaves are withered to the extent that they droop downwards, according to the Chazon Ish the aravah is pasul.³

A withered aravah.

---

3. This is because the aravah then does not resemble "the way it grows," a necessary requirement for all of the Four Species. Rav Chaim Kanievsky encouraged the author to publicize this halacha.

The leaves are withered and droop downward.

4. Today, since we are blessed with an abundance of easily attainable aravos, it is proper to replace the aravos daily with fresh ones.

An Aravah with a Cut-Off Top

1. If the top of the aravah's branch is cut off, it is pasul for the first day of Yom Tov according to all opinions. For the rest of the days of Sukkos, the Mechaber holds that *l'chat'chilah* it is preferable not to use it, but *b'dieved* if he has no other aravah, it is kosher and, on Chol HaMoed, a *berachah* may even be recited over it. The Rema however holds that it is pasul for the entire duration of Sukkos.

Whereas Sephardic Jews generally follow the opinion of the Rema, the Kaf HaChaim rules that in this, it is best for Sephardim as well to follow the more stringent opinion of the Rema.

2. If the leaf at the top of the aravah is cut off, but the branch remains intact, the aravah is kosher.

3. If smaller branches grow from the central branch of the aravah, it is not necessary to remove them. However, removing them does not disqualify the aravah, as would the removal of the top of the aravah's central branch.

The top of the aravah's branch has been cut off.

If the leaf at the top of the aravah is cut, the aravah remains kosher.

Smaller branches growing from the central branch of the aravah.

Lavluv*

1. The term *lavluv* refers to the uppermost leaf that protrudes from the tip of the aravah and often curls around itself, thus enveloping the aravah's tip. As the aravah grows, the *lavluv* straightens out, ultimately taking on the appearance of the other, lower leaves. The branch then continues to grow, forming another *lavluv*.
2. If the aravah does not have a *lavluv*, this could be due to one of a number of reasons:

   a. The aravah was taken from the tree after the *lavluv* had matured and straightened into a standard leaf, before a new *lavluv* was formed.

   b. The *lavluv* was removed from the tip. As long as the tip remains undamaged, the removal of the *lavluv* does not render the aravah pasul.

   c. The tip of the aravah was removed, together with the *lavluv*.

Only in this third and final case is the aravah pasul. However, since it is difficult to ascertain the cause of the *lavluv*'s disappearance, it is best to use an aravah with a *lavluv* in order to ensure that the tip has not been damaged.

---

* The origin of the word *lavluv* is unclear; it may be a colloquial term for the Hebrew *livluv*, which means "new growth."

An aravah with a lavluv at its tip.

An aravah without a lavluv at its tip; it is difficult to discern whether the tip of the aravah has been cut.

An Aravah whose Leaves have Fallen Off

1. An aravah from which the majority of leaves have fallen is pasul. If a count of the stubs of the severed leaves reveals that even a slight majority of leaves remain attached to the branch, the aravah is kosher. It is not necessary for there to be an obvious majority of attached leaves.

The majority of the leaves have fallen off of the aravah.

2. Since aravos are easily attainable, it is best not to use an aravah from which even a minority of leaves has fallen.

Esrog

***Our Sages taught that the "fruit of a hadar tree" mentioned in the verse refers to the esrog, whose fruit has the same taste as the tree itself. [Succah 35a]
(The wording of the passage in Hebrew implies such a quality.)***

A Chazon Ish esrog

A Yemenite esrog

A Moroccan esrog

A Kalabrean esrog

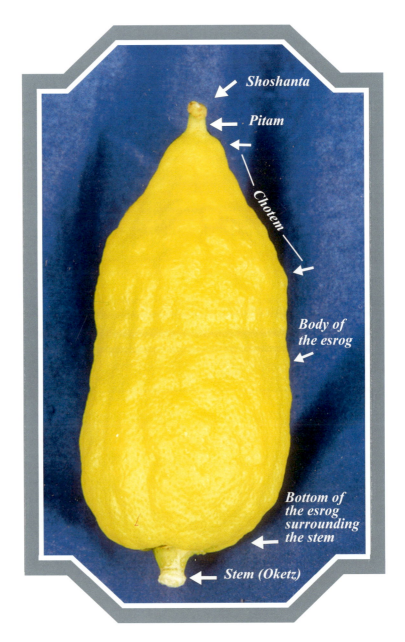

The different parts of the esrog.

Description of the Esrog

The esrog consists of the following areas:

1. At the uppermost tip of the esrog can be found a small protrusion called the *pitam*. The brown incrustation at the tip of the *pitam* is called the *shoshanta*.
2. The upper slope of the esrog, where it begins to get narrower until the actual top, is called the *chotem*. The main body of the esrog is below the *chotem*.
3. At the very bottom of the esrog is its stem (*oketz*), by which the esrog is connected to the tree. The area surrounding the *oketz* is usually recessed.

Terms that Describe Discoloration of the Esrog

Certain discolorations of the esrog may make the esrog pasul, depending on their origin and location. For the sake of clarity, we will categorize them as follows:

1. A "pasul spot" is a discoloration that has the potential of making an esrog pasul. In terms of halacha, it is identical to the blister-like *chazazis* that we will discuss later on. Generally speaking, these spots have either a black or white color.
2. A "kosher spot" is a discoloration that does not make the esrog pasul. However, it can detract from the beauty of the esrog. Some hold that a profusion of kosher spots can render the esrog pasul.
3. Some apparent discolorations are not even spots at all, such as dirt on the esrog.

4. A *bletel* is a scab on the esrog that has formed over natural damage, while the esrog is still attached to the tree. A full discussion of the halachos of *bletelach* (pl.) appear later on.

Size Requirements of the Esrog

1. The minimum size requirement of an esrog is one that displaces 2 fl. oz (57.6 cc) according to Rav Chaim Na'eh, and 3.4 fl. oz (99.5 cc) according to the Chazon Ish when placed in

An esrog weighing 3.5 ounces (100 grams) is well above the minimum requirement.

a full container of water. Alternatively, an esrog can be measured by weight. An esrog weighing 90 grams (3.2 ounces) can be assumed to have a volume of 100 cc, which is kosher according to both opinions. However, to protect against possible future shrinkage (see below), an esrog weighing 100 grams (3.5 ounces) is preferred.

2. An esrog has no maximum size limit; even if it is so large that it must be carried with both hands, it is kosher.

3. An esrog that once fulfilled the minimum size requirement, but shriveled during Chol HaMoed to less than the minimum, is pasul.[1] The same is true for an esrog that was damaged and reduced to less than the minimum size requirement.

4. It is proper to use an esrog that is significantly larger than the minimum requirement, in order to ensure that even if it does shrink, it will still remain kosher.

A Hybrid Esrog

1. An esrog can be crossed with a lemon by grafting the branch of an esrog tree onto a lemon tree or vice-versa, thus strengthening the esrog strain. In either case, the resulting hybrid is pasul.

2. A hybrid esrog that is the result of cross-pollination (i.e. bees carried pollen from lemon flowers to pollinate the esrog) is kosher.[2]

---

1. The Mishnah Berurah in the Bei'ur Halachah rules that it is pasul, whereas the Chazon Ish (148:2) rules that it is kosher.

2. Rav Shlomo Zalman Auerbach *zt"l*, quoted in *Kashrus Arba'as HaMinim*, page 182.

3. All this is where the esrog was grafted with a lemon. However, an esrog that is a hybrid of two different varieties of esrog is kosher.

4. A hybrid esrog can be recognized through the following signs:
   a. A non-hybrid esrog's stem emerges from an indentation in the base, i.e the area surrounding the stem is recessed. In a hybrid esrog there is no such indentation.
   b. The skin of a hybrid esrog is smooth, whereas a non-hybrid esrog's skin is bumpy.
   c. The inner white flesh of a hybrid esrog's peel is thin like a lemon's, whereas a non-hybrid's flesh is thick.
   d. The seeds of a hybrid esrog lie across the width of the fruit, whereas a non-hybrid's seeds lie along its length.

5. These signs are not always reliable, and therefore should not be depended upon for the first day of Yom Tov. Rather, an esrog from a variety that is known by tradition to be kosher must be used. This is similar to the halacha that only birds from a variety traditionally known to be kosher may be eaten, even though the Talmud specifies signs to recognize a kosher bird.

Since there are pasul hybrid esrogim on the market, it is important to buy esrogim that have a certification of kashrus.[3] Esrogim that are homegrown, although the owner claims not to have crossed them, must not be used unless the owner knows for certain that his esrog trees are from a pure variety.

6. For the second day of Yom Tov and the duration of Chol HaMoed, when the *mitzvah* of esrog is *mi-d'Rabbanan*, the

---

3. Mishnah Berurah 648: 65.

Esrog 127

The stem of the top esrog grows from an indentation on the bottom.
The stem of the bottom esrog juts out from the bottom.

An esrog whose stem juts out from its base.

A cross section of an esrog

1. Its skin is bumpy. 2. Its stem grows from an indentation on its bottom.
3. It has thick inner white flesh. 4. Its seeds lie across its length.

Esrog 129

A cross section of a lemon

1. Its skin is smooth.   2. Its stem juts out from its bottom.
3. It has thin inner white flesh.   4. It seeds lie across its width.

above-mentioned signs may be relied upon in cases of great necessity. It is not necessary to cut open the esrog to investigate the nature of its white flesh or seeds; the outer signs of the deep-set stem and the bumpy skin are sufficient to verify the kashrus of the esrog.

7. Furthermore, even on the first day of Yom Tov, if no other esrog is available, or if one already purchased on esrog without certification and cannot exchange it, the outer signs of the deep-set stem and bumpy skin may be relied upon, and a *berachah* may be recited.

8. If only one of the outer signs of kashrus is present, the esrog should not be used. However, if a different esrog from the same tree was cut open and the inner signs of kashrus were found (the thick white flesh and the lengthwise seeds) along with one of the outer signs, the original esrog may be used and a *berachah* may be recited.

9. The chemical pesticides used in esrog cultivation can often affect the properties of the fruit. Therefore, an esrog that is known to have come from a pure variety may be used even if some of its characteristics resemble those of a lemon, as described above.[4]

---

4. Heard from Rav Nissim Karelitz.

An esrog with smooth skin and a stem that juts out from its bottom.

10. An esrog that is known to be hybrid is pasul for the entire duration of Sukkos. If no other esrog is available, some opinions hold that a hybrid one should be used (without a *berachah*) in order to perpetuate the mitzvah of lulav and esrog for future years. Some opinions hold that it should not be used, in order that one not become accustomed to using such a pasul esrog in future years, when perhaps a kosher one would be available.

Chazazis

Our Sages discuss a bumpy growth called a *chazazis* that may render the esrog pasul, depending on its size and location. The *poskim* disagree as to the exact definition of the *chazazis* discussed. Some define it to be a fluid-filled bubble that develops on the skin of the esrog as a result of infection, much like a blister that develops on human skin due to a burn injury. Some define it to be an unusual, needle-like formation of the skin of the esrog.[5]

1. Even a minuscule *chazazis* found on the *chotem* of the esrog renders it pasul, provided that it is visible to the naked eye.[6]

2. A *chazazis* on the rest of the esrog will render it pasul only if it covers the greater part of the esrog's surface.

---

5. Heard from R' Eliezer Daube, an esrog grower, who invited various *poskim* to inspect the esrogim in his orchard. When asked to point out the *chazazis* discussed in halacha, they offered differing replies, as detailed above.

6. This does not include the *pitam* or *shoshanta*. We treat this topic further in our discussion of the *pitam* later on.

3. According to the Mechaber, two separate *chazaziyos* found on the main body of the esrog automatically render the esrog pasul.[7] However, according to the Rema, an esrog with two *chazaziyos* is pasul only if, when looked at superficially, the *chazaziyos* appear equidistant whichever way one views them.

4. An esrog with three or more *chazaziyos* that spread out over the majority of the circumference of the esrog is pasul according to all opinions.

5. A *chazazis* will invalidate the esrog only if it rises above the surface of the esrog to such a degree that the hand clearly feels the protrusion. If the *chazazis* feels like mere roughness but not a noticeable protrusion, the esrog is kosher.

6. A bump of unusual size or shape does not render the esrog pasul, provided that it is not in the category of *chazazis* described above in the introduction to this section.[8]

7. If the reason the *chazazis* rises above the surface of the esrog is only because it rests on a natural bump of the esrog, the esrog is kosher.[9]

---

7. Kaf HaChaim 648:74.

8. Heard from Rav Chaim Kanievsky, Rav Ezriel Auerbach and Rav Nissim Karelitz.

9. Heard from Rav Refoel Reichman.

An esrog with an unusually large bump on its surface. Such an esrog is kosher.

An esrog with an unusually large bump on its surface. Such an esrog is kosher.

*An esrog with an unusually large bump on its surface.
Such an esrog is kosher.*

A Dried-Out Esrog

1. A dried-out esrog is pasul. Some opinions interpret "dried-out" to mean that the greater part of the esrog is dried-out. Some hold that even if a small portion of the esrog is dried-out, it is still pasul. Others hold that the dry parts of an esrog invalidate the esrog in the same manner as *chazazis* does (see above).

2. The definition of dryness is that when a threaded needle is drawn through the esrog, no moisture is felt on the thread. Care should be taken when performing such a test not to puncture the middle, seedy part of the esrog. As we will soon see, such a hole would adversely affect the kashrus of the esrog. Rather, the needle should be drawn lengthwise through the flesh of the esrog's side that curves away from its center. Alternatively, the eye of the threaded needle may be inserted into the esrog and removed, thus eliminating the necessity to draw the entire length of the needle through the esrog.

3. An esrog left over from the previous year is certainly dried-out and pasul. Even if a needle and thread test would find moisture in the esrog, it would nevertheless be pasul. However, if the esrog was properly preserved to retain its freshness, then a needle and thread test may be performed and relied upon to ensure the moistness and kashrus of the esrog. In this case, even if the esrog would appear to be fresh, a needle and thread test would be necessary to verify its moistness.

4. An esrog from the current year that does not show any apparent signs of dryness need not be tested for moistness.

A dried-out esrog.

A Punctured Esrog

1. In a case where the esrog is punctured, yet no part of the esrog is missing, the halacha is as follows: If the puncture runs widthwise through the entire esrog, from one side to the other, the esrog is pasul. If the puncture does not extend to the other side of the esrog, the esrog is kosher and a *berachah* may be recited over it.

If the puncture reaches until the seedy area of the esrog, some *poskim* hold that the esrog is pasul, even if the hole does not extend to the other side and the puncture runs lengthwise. If at all possible, such an esrog should not be used. However, if the only esrog available has such a puncture, then it may be used and a *berachah* may be recited over it.

A needle that has been drawn widthwise through the esrog.

2. If it is unclear whether the puncture reaches the seedy area, the esrog is considered kosher. This phenomenon is common with Yemenite esrogim, which often have a hole at their top, and it is difficult to ascertain whether this hole descends to the seedy area.

An esrog with a hole on its tip, that might possibly descend into the seedy area.

3. All this is in regard to an esrog with a puncture that runs through its width. If the puncture runs through its length, even if it extends through the entirety of the esrog, the esrog is certainly kosher provided that the puncture does not enter the seedy area.

A needle that has been drawn lengthwise through the bumps of the esrog.

4. If the puncture does not run through the body of the esrog, but rather through one of the bumpy protrusions on the skin of the esrog, even if the puncture runs widthwise, the esrog remains kosher.

5. Even if the puncture does not run through the entirety of the esrog, if it is ¾ in. (nineteen millimeters) or larger in diameter, the esrog is pasul.

An esrog with a hole ¾ of an inch in diameter.

6. In any of the cases discussed above, in which the puncture does not invalidate the esrog, the esrog should be inspected to make sure that no part of it was removed in the process. The halachos of an incomplete esrog will be discussed below.

The Midrash compares the Four Species to the four categories of Jews. When we bind the Four Species together, it is symbolic of binding the entire Jewish people together, allowing each one's advantages to compensate for the other's faults.

Why did the Torah command us to perform a symbolic act? Why did the Torah not command us to actually bind together as a nation, and unite the righteous with the wicked? Perhaps the interaction of the tzaddik and the rasha might have the unfortunate consequence that the tzaddik will be influenced by the rasha's wicked ways, and not the opposite. Therefore, the Torah commanded us to bind the Four Species instead. This symbolic act could affect a beneficial union for the Jewish people on a spiritual plane, without any harmful side-effects.

(The Satmar Rav)

An Incomplete Esrog

1. If even a small portion of the esrog was actually removed, the esrog is pasul for the first day of Yom Tov, but kosher for the days of Chol HaMoed. For the sake of brevity, such an esrog will hereafter be referred to as damaged.

2. Although the second day of Yom Tov outside of *Eretz Yisrael* generally has the same stringencies as the first day,[10] in regards to a damaged esrog this is not the case. Whereas on the first day of Yom Tov even a slightly damaged esrog is pasul, on the second day it is kosher, unless a hole the size of ¾ in. (nineteen millimeters) in diameter or more is missing.

3. Even on the first day of Yom Tov, if the only esrog available is one that is damaged by a hole not larger than ¾ in. (nineteen millimeters) in diameter, it is kosher and a *berachah* may be recited over it according to the opinion of the Rema. According to the Mechaber, however, a *berachah* may not be recited over it on the first day.[11]

4. If a superficial inspection of the esrog does not reveal any signs of damage, it is unnecessary to check the esrog with a magnifying glass to ensure that it is indeed undamaged. Even if such an apparently complete esrog were to reveal a tiny imperfection upon inspection with a magnifying glass, the esrog would still be kosher. The Torah was not given with the expectation that the Jewish people would need to use magnifying glasses to ensure the fulfillment of its mitzvos.

---

10. However, any of the *arba'ah minim* that is pasul for the first day of Yom Tov may be taken on the second day, without a *berachah*, if it is difficult to find a substitute (Shulchan Aruch 649:5).

11. Heard from Rav Refoel Reichman.

5. If it is unclear whether or not the esrog is damaged, then it is kosher, and a *berachah* may be recited over it. Very often, what is assumed to be real damage is in fact questionable. Sometimes the flesh of the esrog is pressed inward giving the impression of damage, but none of the esrog was actually removed. In such cases of uncertainty, the esrog is considered kosher, but not *mehudar*.

6. If it is unclear whether or not the esrog is damaged upon inspection with the naked eye, and closer inspection with a magnifying glass reveals it to be undamaged, it is not considered to be in the category of questionably damaged, and it is classified as *mehudar*.[12]

7. An esrog that was damaged by thorns or by a person while it was still attached to the tree, and later healed itself by growing an incrustation over the hole, is kosher. If the incrustation was not completed before the esrog was picked, and a portion of the hole is still apparent, the esrog must be judged according to the guidelines discussed above concerning punctured or damaged esrogim.

8. If it is uncertain whether the thorn actually removed part of the esrog, making it pasul, or merely punctured it leaving it still kosher, then the matter must be decided according to the superficial appearance of the esrog. If the esrog appears to be incomplete, it is pasul. If it is questionable, the esrog is kosher, since thorns generally do not remove part of the esrog.

---

12. Heard from Rav Moshe Shapira.

An esrog punctured by a thorn.

9. The distinction between a damaged, pasul esrog, and a questionably damaged, kosher esrog, depends upon its appearance to the naked eye.

10. A *bletel* is a scab on the esrog that formed over natural damage, while the esrog was still attached to the tree. If a *bletel* on the esrog has cracked, and beneath it the green skin of the esrog is visible, if the skin has a shiny luster, then it may be assumed that the esrog was damaged and completely healed itself before it was picked from the tree. Therefore it is kosher.

If, however, the green skin beneath it is not shiny, then it is questionable whether the damage was completely healed while the esrog was still on the tree; therefore its status is that of a questionably damaged esrog, as described above.

The halachos of *bletelach* will be discussed later in greater detail.

12. A distinction must be made between damage to the white, inner flesh of the esrog, and its green or yellow outer skin. See below in the section discussing a peeled esrog.

13. An esrog with an indentation that is not due to damage, but rather is the natural shape of the esrog, is kosher. If the in-

An esrog with an unusually deep indentation.

dentation is readily noticeable, the esrog might be considered unsightly, and perhaps should be avoided in favor of a more beautiful esrog for the sake of *hiddur mitzvah*. In any case, the esrog remains kosher.[13]

A Crushed Esrog

L'chat'chilah it is better not to use an esrog whose inner white flesh is crushed, even if the outer green or yellow skin is intact. *B'dieved*, it is kosher. Such an esrog is readily detectable when held.

A Cracked Esrog

1. If a sufficient part of the esrog (see below) is cracked, and the crack descends into the inner white flesh of the esrog, even if the crack is only visible from one side, the esrog is pasul.

2. According to some opinions, if the crack does not extend for the entire length of the esrog from the *pitam* to the stem, the esrog is kosher. According to other opinions, if the crack reaches even slightly into the *chotem* (the upper slope of the esrog), the esrog is pasul. Some opinions hold that even if the *chotem* is entirely intact, if the crack extends for the greater part of the length of the esrog, the esrog is pasul.

3. The question of whether a crack is sufficiently large to invalidate an esrog applies only where the crack does not reach the seedy area of the esrog. If the crack descends into the seedy area, however, we must consider it according to the guidelines

---

13. Heard from Rav Chaim Kanievsky.

of a punctured esrog described above.

A Peeled Esrog

1. An esrog has three layers:
   a. The outermost shiny covering, known as the cuticle.
   b. The sharp-tasting green or yellow skin underneath the cuticle. This skin contains a fragrant juice that gives the esrog its distinctive smell.
   c. The white, innermost flesh of the esrog. Within this fleshy part is a circular pocket containing the esrog's seeds.

2. According to some opinions, if the entire thickness of the green skin has been peeled from part of the esrog revealing the white flesh, but the white flesh has not been damaged at all, the esrog is not considered incomplete and it is kosher.

Others hold that only if part of the thickness of the green skin has been peeled, and enough of the skin remains intact to obscure the white flesh, is the esrog kosher. However, if the entire thickness of the green skin is peeled, revealing the white flesh, the esrog is considered damaged and pasul.

The accepted halacha is to be stringent in accordance with this second opinion. However, we have already stated that a damaged esrog is kosher for the duration of Chol HaMoed.

3. Even in a case where the esrog is peeled but not damaged, if the peel has been removed from the entire esrog, it is pasul. Some hold that if any portion of the peel remains, the esrog is kosher, while others hold that for the esrog to be kosher the remaining portion of skin must be at least the size of a *selah*.[14]

---

14. A *selah* was an ancient coin. It was approximately 1 inch (25 millimeters) in diameter (*Middos Umishkalos shel Torah*, p. 367).

4. If the place where the esrog was peeled develops a black or white color, then the esrog has the same status as if it were spotted with pasul spots. (An esrog with pasul spots is discussed later on.)

The place where the esrog was peeled developed a black color.

We will see that there is a difference of opinion among *poskim* as to whether or not a brown spot renders the esrog *pasul*. In cases of necessity, if a brown color appears where the esrog was peeled, the lenient opinions may be relied upon and the esrog is considered kosher.

> *There is a custom of making preserves from the esrog, to be eaten on Tu b'Shevat. This is said to be a segulah for women, that they have easier births and healthy children.*
>
> *(Kaf HaChaim 664:9)*

The *Pitam*

1. The *pitam* is a protrusion emerging from the tip of the esrog's upper slope, or *chotem*. Surrounding the head of the *pitam* is a brown, crown-like wreath, known as the *shoshanta*.

There are two kinds of *pitam*: fleshy and woody. The fleshy *pitam* appears in all respects like the esrog itself, with the same green or yellow color as the esrog's skin, and the same white interior. Therefore, some hold the opinion that the various halachic stringencies regarding the *chotem* of the esrog itself apply to the fleshy *pitam* as well. For example, if there is a pasul spot on the fleshy *pitam*, or the *pitam* is damaged even slightly, the esrog is pasul. However, the woody *pitam* is not considered part of the esrog and therefore these stringencies do not apply.[15]

Other opinions are lenient in regard to the fleshy *pitam* as well.

An esrog with a green, fleshy pitam.

---

15. Kaf HaChaim 648:46.

A brown, woody pitam on a fully developed esrog.

A brown, woody pitam on an undeveloped esrog.

2. If the *pitam* fell off the esrog while the esrog was still attached to the tree, even moments before it was picked, according to some opinions the esrog remains kosher.[16]

---

16. Heard from Rav Chaim Kanievsky, in the name of the Chazon Ish. This is also the opinion of Rav Shlomo Zalman Auerbach *zt"l*, as quoted in *Kashrus Arba'as HaMinim*.

Esrog 155

An esrog whose pitam fell off while still attached to the tree.

Furthermore, if while still attached to the tree, the *pitam* had dried up to the extent that it no longer remained firmly attached, this is also considered as if it had fallen off. If the *pitam* falls off by itself after the esrog was picked from the tree, then it can be assumed that it had already dried before while on the tree, and the esrog is kosher.[17] However, if after being picked, its *pitam* is broken off due to external pressure, then it must be assumed

---

17. Heard from Rav Chaim Kanievsky.

that the *pitam* was still firmly attached before, and therefore the esrog is pasul.

A pitam that withered while still attached to the tree, yet fell off only after the esrog was picked.

A clear indication that the *pitam* was fully dried while still on the tree is the presence of a scab that fully covers the place where the *pitam* was severed. Therefore, whether the *pitam* fell off by itself or was broken off, it is necessary to check the tip of the *chotem* for the aforementioned scab, to ensure that the esrog is kosher. If the place of the *pitam* is only partially covered with a scab, then it must be assumed that it was still partially attached while on the tree, and only fell off later, after the esrog was picked; consequently, the esrog must be considered pasul.

Black spots found on the scab do not render the esrog pasul.[18]

3. All of the above follows the opinion that a *pitam* that fell off even immediately before the esrog was picked does not render the esrog pasul. However, others hold the opinion that only if the *pitam* fell off in the early stages of the esrog's growth, does the esrog remain kosher.[19] If there is an indentation at the tip of the *chotem*, where the *pitam* was once attached, this is a clear

An indentation on the esrog's chotem that developed a scab.

---

18. Heard from Rav Chaim Kanievsky.

19. Heard from Rav Ezriel Auerbach, in the name of Rav Yosef Shalom Elyashiv.

indication that the *pitam* fell off in the early stages of the esrog's development, and the esrog is kosher. Another indication that the *pitam* fell off in the early stages of the esrog's growth is a thin groove that encircles the area of the fallen *pitam*.

4. In cases of uncertainty, wherein the aforementioned signs fail to clearly discern when the *pitam* fell off, the esrog may not be used for the first day of Yom Tov, but may be used for the other days of Sukkos, and a *berachah* may be recited over it.[20] However, if it is certain that a fleshy *pitam* fell off after it was picked, the esrog should not be used for the entire duration of Sukkos.

> *The Midrash relates each one of the arba'ah minim to Hashem Himself. What our Sages mean to tell us by this is that whereas each species in creation has a ministering angel overseeing its growth, the arba'ah minim have no such ministering angel – Hashem Himself oversees their growth.*
>
> *(Alshich)*

---

20. This is due to a combination of leniencies: 1. Perhaps the *pitam* fell off while the esrog was still attached to the tree. 2. Even if the *pitam* fell off afterwards, perhaps this is only considered a damaged esrog, which is kosher for Chol HaMoed. 3. Even if it is considered a deficiency in *hadar*, there are those opinions that hold that *hadar* does not apply to Chol HaMoed. (We previously discussed the *hadar* requirement by the laws pertaining to lulav.)

A *berachah* may be recited on the second day of Yom Tov, as well. See Bikurei Ya'akov 649:32.

The Growth of the *Pitam*

There are two seasons in which the flowers of the esrog tree begin to bloom: in the early springtime months of Adar and Nisan, and the early summer months of Sivan and Tammuz. Even in these months, the flowers do not bloom all at once, but rather one at a time over the course of the season.

Thus, when the Gemara says that the esrog fruit grows on the tree throughout the year, this does not mean to say that the tree blossoms and fruit begins to grow throughout the year. Rather, the fruit that has grown in its season can remain attached to the tree and continue to grow throughout the year, and sometimes for more than one year as well.

Some esrog growers utilize both growing seasons, whereas others prefer not to use the esrogim grown in the spring. It is

Esrogim growing during the flowering season.

difficult to preserve such esrogim for so many months before Sukkos. Furthermore, if the esrog is left on the tree to grow, it becomes oversized and unsightly, and thus difficult to market. Therefore, in order to eliminate the first season's esrogim, these farmers refrain from spraying the trees with pesticides, as they otherwise would. The insects then consume the budding fruit. In areas where insects do not infest the crop, the esrog growers must go through the trouble of removing the first season's fruit themselves, in order to enhance the growth of the second season's fruit.

The small flowers that eventually develop into esrogim are pink outside and white inside. Not all flowers have the capacity to develop into fruit. Those that do have a pistil, a small sticklike growth inside. The yellow top half of the pistil eventually develops into the *pitam*, and the green bottom half develops into the fruit of the esrog. Surrounding the pistil grow yellow stamen. Insects transfer pollen from the stamen to the pistil, enabling the esrog to develop.

Flowers that have not yet opened.   *A flower that will develop into an esrog.*

The pistil of the esrog flower. The top, yellow half is the pitam, and the bottom, green half will eventually develop into the fruit.

In the early stages of the esrog's development the *pitam* begins to crack and separate from the fruit. The crack is visible as a ring that surrounds the *pitam*, and becomes deeper and deeper until the *pitam* finally falls off. Most *pitam*s fall off as a natural process of the esrog's growth. In order to strengthen the *pitam* so that it remains attached to the fruit, those esrog growers who desire esrogim with a *pitam* spray them with pesticides after the stamen has been pollinated and the petals have fallen off, yet

162 ◈ THE FOUR MINIM

An undeveloped esrog whose pitam is beginning to crack.

before the *pitam* has begun to crack. If the *pitam* is sprayed after it has already begun to crack, the part that is cracked develops a scab. Sometimes this scab can be felt as a protrusion above the surface of the esrog. Even so, this does not detract from the kashrus of the esrog. Even the Maharil who is stringent in regard to a *bletel* that is felt above the surface of the esrog (see our dis-

A pitam that has fallen off of the esrog.

cussion of *bletelach* later on), agrees that the scab on the *pitam* is kosher.[21]

The reason for this leniency is that the scab does not repre-

---

21. Heard from Rav Chaim Kanievsky and Rav Nissim Karelitz.

A scab on the pitam.

sent a deficiency or disease of the esrog. The natural process of the esrog's development would have severed the *pitam* entirely if not for the grower's intervention. Therefore the marks of the crack that remain are kosher.

On the other hand, the Maharil considers an upraised *bletel* pasul because it is the result of damage to the esrog. The esrog then heals itself by forming the *bletel* scab. This is the mark of an unnatural injury to the esrog, and is therefore similar to one caused by disease (*chazazis*) that would render the esrog pasul.

Rav Moshe Heller *zt"l* added that *bletelach* found on the *pitam*, whatever their origin may be, do not render the esrog pasul. However, if the *bletel* is unsightly then the esrog may be lacking in the quality of *hiddur mitzvah*, i.e. performing a mitzvah in a beautiful manner.

During the Shemittah year in *Eretz Yisrael*, when the growers are forbidden to spray the *pitam* to strengthen it, it is unlikely to find the circular scab around the *pitam* described above. However, there are rare cases in which an esrog that has been well watered has the strength to preserve its *pitam* although it has begun to crack. In such cases the aforementioned scab would form.

Pesticides

In Israeli esrog orchards a certain pesticide is sometimes used, which is then absorbed by the esrog, rendering it poisonous and unfit for consumption. As we will see below, an inedible esrog is pasul, and therefore this pesticide should not be used after the 15th of Tammuz, three full months before Sukkos. After three months no traces of the pesticide remain in the esrogim.

A *Pitam* that Fell Off

1. An esrog with an undamaged woody *pitam* whose *shoshanta*

fell off before it was picked from the tree, whether by itself or as a result of external pressure, is kosher *l'chat'chilah* according to the opinion of the Mechaber. According to the Rema, however, if an esrog with an undamaged *shoshanta* is easily obtainable, it should be used instead. If not, though, then this one may be used and a *berachah* may be recited over it.

2. A woody *pitam* that fell off, leaving an indentation in its place at the tip of the *chotem*, is considered damaged and pasul for the first day of Yom Tov.[22] If no other esrog is available, it may be used, however, and a *berachah* may be recited.

3. If some of the woody *pitam* fell off with the *shoshanta*, leaving a portion of the *pitam* still attached, and that portion rises above the surface of the hole in which it is embedded, the esrog is kosher *l'chat'chilah* according to the opinion of the Mechaber. According to the Rema, it should not be used *l'chat'chilah*. If no other esrog is available, however, it may be used and a *berachah* may be recited.

4. If the remaining portion of the woody *pitam* does not rise above the hole, but fills it entirely, making it level to the surrounding area of the esrog, the esrog is considered pasul for the first day of Yom Tov, but kosher for the rest of Sukkos.[23]

5. An esrog with a fleshy *pitam* that fell off after the esrog was picked from the tree, is pasul for the entire duration of Sukkos.

---

22. It is kosher for the second day of Yom Tov because the hole will certainly be less than ¾ in. (nineteen millimeters) in diameter (Mishna Berurah 648:8).
23. See Bikurei Ya'akov 649:32.

If it is unclear whether the *pitam* fell off before or after the esrog was picked from the tree, the esrog is pasul for the first day of

*On the right – an esrog whose pitam fell off while still attached to the tree.
On the left – an esrog whose pitam fell off after it was picked.*

Yom Tov, but may be used for the rest of Sukkos, and a *berachah* may be recited.

6. If the *pitam* was broken but not completely severed, some *poskim* hold that if it is still firmly attached to the esrog, the

esrog is kosher. Others hold that the esrog is pasul regardless of whether the broken *pitam* remains firmly attached or not.

An esrog whose pitam is broken yet still attached.

Rebbe Levi Yitzchak of Berditchev's great love of mitzvos was legendary. Many days before the arrival of Sukkos, he was overcome with longing to fulfill the mitzvah of arba'ah minim. No price was too high for him to pay for an esrog, and no obstacle too difficult for him to overcome.

When he finally acquired a beautiful, mehudar esrog, he gently wrapped it in flax, and tucked it away for safekeeping. When the morning of Sukkos finally arrived, Rebbe Levi Yizchak joyously made his way to the mikveh to prepare himself for the Yom Tov prayers and for his beloved mitzvah of arba'ah minim. While he was gone, his assistant awaited his return, and thought to himself: only after the Rebbe, and the entire congregation had already recited the blessing over the esrog, would he finally have the opportunity to do so himself. Perhaps he could take the esrog now, while the Rebbe immersed in the mikveh. Thus he would be the first to perform the mitzvah!

With awe and trepidation, he slowly opened the esrog box. His hands trembled as he unwound the flax that surrounded the esrog, and then...alas! The esrog fell and its pitam broke.

The assistant was beside himself with anguish and regret. How could he have had the audacity to take his Rebbe's esrog without permission?! Through his own impatience and selfishness, the Rebbe would now be lacking the arba'ah minim. How would he explain to the Rebbe what he had done?

When the Rebbe returned from the mikveh, he went to recite the berachah and perform his beloved mitzvah. Shamefully, the assistant explained to the Rebbe what he had done. The assistant studied his Rebbe's face, waiting to see even the slightest hint of anger. To his great surprise, however, the Rebbe's face shone with joy! The Rebbe turned to the Heavens and proclaimed, "Master of the universe! Who can compare to Your precious nation, Yisrael! Behold — a simple Jew, willing to risk ruining my esrog out of his great eagerness to perform Your mitzvos!"

Unsightly Esrogim

1. As we explained by the laws of lulav, we are obligated to perform mitzvos in a beautiful way — *hiddur mitzvah*. An esrog with a peculiar shape is considered unsightly. Therefore, although it is kosher[24], it should not be used, for it is missing the quality of *hiddur mitzvah*.

An esrog with an unusual and unsightly appearance.

---

24. Heard from Rav Chaim Kanievsky, who explained that any deformation not explicitly discussed by the *poskim* can be assumed to be kosher.

2. Esrogim with the following characteristics are considered peculiar, and should be avoided, if more beautiful esrogim are available:

   a. The esrog is thin at its stem, and progressively gets wider towards the *pitam*. (Such a phenomenon is not uncommon among the variety of esrogim called Yemenite esrogim.)

   b. The stem of the esrog and its tip are not aligned. If the misalignment is only slight, the esrog is still considered *mehudar*. If it is severe, the esrog is considered kosher but not *mehudar*. Even so, a misaligned esrog is considered more beautiful than one that gets wider towards the *pitam*.

On the right – an esrog whose pitam is not aligned with its stem.
On the left – an esrog that is thin at its stem,
and progressively gets wider towards the pitam.

*An esrog whose pitam is considerably misaligned with its stem.
It is best not to use such an esrog.*

Our Sages tell us that the esrog is unique in that it remains attached to the tree from year to year. Perhaps we can find a lesson in this. Many people have the custom of changing their behavior from Rosh Hashanah through Sukkos. They pray fervently and are scrupulous in their observance, out of fear and trepidation of the Day of Judgment. As soon as Sukkos passes, however, they return to the same routine as before Rosh Hashanah. The esrog represents the tzaddik, who is not like this. He remains attached to the Tree of Life for the entire year, from one Rosh Hashanah to the next.

(Pardes Yosef, parashas Noach)

The Stem

It is interesting to note that the variety of an esrog can be determined by its scent when the stem is cut. The Ording and Braverman esrogim (also known as Maharil Diskin esrogim) give off a sharp lemon smell, whereas the Chazon Ish esrogim do not.[25]

The stem (oketz) that attaches the esrog to the tree.

1. If the stem of the esrog has been removed, leaving an indentation, the esrog is considered damaged and is pasul for the first

---

25. Heard from Reb Avraham Ludmir, a noted esrog grower.

day of Yom Tov, but kosher for the other days of Sukkos.[26]

If no other esrog is available, such an esrog may be used even on the first day, and a *berachah* may be recited.

2. However, if the stem has not been removed, but only cut, so that it is still connected to the esrog and the indentation within which it resides has not been exposed, the esrog remains kosher, even for the first day of Yom Tov. This is so, even if the remaining part of the stem only fills the indentation and does not rise above it at all.

The Origins of Black Spots

There are six distinct types of black spots that can be found on the skin of the esrog. If the black spot is a discoloration of the esrog itself, it may render the esrog pasul, depending on its size and location. If the black spot is merely dirt that lies upon the esrog, the esrog remains kosher. Any spot that is readily removable without damaging the esrog, is clearly not part of the esrog itself, and the esrog is kosher.

1. Some black spots are a result of the pesticides used to protect the esrog from the more than seventy different kinds of insect infestation that are common among citrus fruits. The pesticides may mix with dust in the air, and settle on the esrog, forming black spots.

2. The pesticides may also leave irremovable ring-like stains that render the esrog pasul. Care must be taken when purchasing

---

26. See note 22. Although the esrog is kosher for the rest of Sukkos, it is still considered of peculiar appearance, and therefore is better not used, as mentioned above regarding *hiddur mitzvah*. (Heard from Rav Nissim Karelitz)

green esrogim to check for these rings. Often they are not clearly visible until after the esrog has turned yellow.

A yellow esrog with marks left by pesticides.

A green esrog with marks left by pesticides.

178 THE FOUR MINIM

3. Some black spots are the result of exhaust particles produced by machines that operate in the vicinity of the esrog orchard. Since these spots are superficial they do not render the esrog pasul.

4. Some black spots are the result of fungus that can grow after the esrog has been picked and packaged. When the fungus marks are found on the outer edges of the esrog's bumpy skin,

An esrog with black fungus.

they may be removed by using modelling clay or any similar adhesive material to lift them off. Such a procedure ensures that no damage is done to the esrog while removing the spots. However, when they are found within the crevices of the bumps, it is often necessary to rub with a toothpick in order to remove them. Extreme care must be taken not to damage the skin of the esrog in the process.

5. There is a certain kind of black spot that is assumed to be the result of insect excretions. This theory has not been fully verified.

6. A sixth form of black spot is of unclear origin, and defies the numerous attempts of experts to ascertain its exact cause.

The Halachos of Black Spots Caused by Discoloration

1. A black spot will invalidate an esrog only if it is clearly visible at a superficial glance. The esrog must be held at the distance one normally holds a book or a watch, approximately 12 inches (30 centimeters).[27] Then the esrog must be turned at a moderate speed. If no black spots are apparent, the esrog is kosher. It is unnecessary to scrutinize the esrog by holding it directly next to the eye, or turning it slowly to meticulously inspect it.[28]

2. If a black spot was first noticed under meticulous inspection,

---

27. Rav Ezriel Auerbach said in the name of Rav Yosef Shalom Elyashiv that when inspecting the esrog it should be held at the same distance as when using it in the performance of the *mitzvah*, which is approximately 30 cm. Some say that the esrog should be checked at a distance of 50 cm (20 inches).

28. Rav Moshe Heller *zt"l* held that if the black spot goes unnoticed the first time the esrog is checked, and later upon rechecking it the black spot is noticed, this is not considered obviously visible and the esrog is kosher.

or through the use of a magnifying glass, and once noticed, it is clearly visible, the esrog is still kosher. The reason for this is that the untrained eye, which has not yet noticed the spot, would not clearly see it.

3. The halachos of the pasul black spot are similar to those of the blister-like *chazazis*, which we discussed earlier. Consequently, a black spot found on the *chotem* of the esrog renders it pasul.

A black spot on the chotem of the esrog.

4. Black spots found on the *shoshanta* do not render the esrog pasul. There is a difference of opinion concerning black spots on the fleshy *pitam*, however, and such an esrog should not be used.

A black spot on the shoshanta.

A black spot on the pitam.

5. An esrog with a black spot on the scab covering the place where the *pitam* was severed is kosher.²⁹

A black spot on the scab that grew over the place of the pitam.

---

29. Heard from Rav Chaim Kanievsky. This is because the scab is considered to be wood, and wood may have a black spot.

6. A black spot on the rest of the esrog will render it pasul only if it covers the greater part of the esrog's surface. According to the Mechaber, two separate black spots found on the main body of the esrog automatically render the esrog pasul. However, according to the Rema, an esrog with two black spots is pasul only if, when looked at superficially, the black spots appear equidistant whichever way one views them. An esrog with three or more black spots that spread out over the majority of the circumference of the esrog, is pasul according to all opinions.

7. Until now we have mentioned only black spots. We will soon see that other colored spots may render the esrog pasul as well. Two or more different colored pasul spots may jointly invalidate an esrog, just as if they all were the same color.

8. Black spots found on the bottommost part of the esrog, surrounding the stem, do not render the esrog pasul. This is because such spots are not readily visible at first glance, as discussed above.[30]

9. An esrog with black spots found on the stem itself is also not pasul.

---

30. Heard from Rav Nissim Karelitz, Rav Ezriel Auerbach, and Rav Yehuda Adess. See also Aishel Avraham (Butchatch) 648.

Esrog 185

Black spots around the stem of the esrog.

Black spots on the stem of the esrog.

10. As discussed later, an esrog with brown spots is generally kosher. However, dark brown spots that appear at first glance to be possibly black must be considered the same as pasul black spots. Once again, this is based on the principle discussed above that the kashrus of the esrog depends on its appearance at a superficial glance.[31]

11. A *bletel* on the *chotem* with a black spot inside renders the esrog pasul.

A black spot on a bletel.

---

31. *Chaim u'Verachah* 648. Also heard from Rav Nissim Karelitz.

12. If there are a few black spots on the *chotem*, although each one might have gone unnoticed by itself and therefore not render the esrog pasul, since the concentration of black spots makes them noticeable, therefore the esrog is pasul.[32]

An accumulation of black spots on the chotem.

---

32. Heard from Rav Nissim Karelitz.

The following story is told of the Chafetz Chaim's outstanding sensitivity to the feelings of others:

During the First World War, the Chafetz Chaim, together with his Yeshiva, were forced to flee their town to Senasvik, Russia. As the Yom Tov of Sukkos approached, the local Jewish community searched far and wide to procure a set of arba'ah minim, but they were unsuccessful.

Finally, someone found a lulav and esrog that had maintained their moistness from the previous year and were fit for use. He hurried to inform the Chafetz Chaim of the good news and present him with the precious find. The Chafetz Chaim was overjoyed. He recited the beracha and held the Four Species, but did not shake the lulav in all four directions.

His closest students were hard pressed to understand why their Rebbe refrained from performing the mitzvah of na'anuim, as prescribed in the Talmud and the Shulchan Aruch.

The Chafetz Chaim explained: "Every year, Jews are accustomed to perform the mitzvah of na'anuim on the arba'ah minim. This year, when the rest of the community is unable to perform the na'anuim, I fear that if they see me shake the lulav it will cause them unnecessary grief. Therefore I preferred to forego the na'anuim, which are merely a minhag, in order not to upset another person, an issur d'Oraysa (Biblical prohibition)."

(Ma'asaih Lamelech)

The Origins of White Spots

There are three different kinds of white spots that can be found on the esrog.

1. White spots may result from the residue of pesticides that were not properly mixed.
2. White discolorations might be due to insufficient sunlight, caused when leaves or other parts of the tree cover a portion of the esrog.
3. White spots might also be insects. The halachos of such infestations will be discussed in a later section.

White discoloration on the bottom of the esrog caused by obstruction of the sunlight.

White discoloration on the chotem of the esrog caused by obstruction of the sunlight.

The Halachos of White Spots Caused by Discoloration

1. The halachos of white spots are identical to the halachos of black spots stated above.
2. A white spot will invalidate the esrog only if it is a discoloration of the esrog itself, and not a foreign object attached to it. Therefore, any white substance that can be removed need not be removed in order to render the esrog kosher. However, practically speaking, it is sometimes necessary to remove the white spot with water or modeling clay in order to ascertain that it is indeed a foreign object.

In cases of uncertainty, we must assume that any white spot is a discoloration of the esrog, until proven to be a foreign object.

3. Almost all of the light-colored spots found and assumed to be white, are in fact pale gray and therefore kosher. The likelihood of finding a truly white spot on an esrog is negligible.[33]
This leniency includes the vast majority of pale *bletelach* that are not fully white, but pale gray.[34]

4. A white spot on an esrog that developed due to the esrog's being obscured from sunlight during growth is considered a pasul white spot. However, if such a spot is only yellow, and not truly white, it is kosher.

---

33. Heard from Rav Ezriel Auerbach in the name of Rav Yosef Shalom Elyashiv.
34. *Orchos Rabeinu*, based on the opinion of the Steipler Gaon *zt"l*.

The Origins and Halachos of Brown Spots

There are ten distinct kinds of brown spots that may be found on the esrog. The halachos pertaining to these spots differ depending on the origin of the spots:

1. Brown spots may be the result of the esrog's being damaged while still attached to the tree. The spot is actually a scab that develops over the damaged part of the peel, and it does not affect the kashrus of the esrog. These scabs are not considered discolorations at all, but rather the natural color of the esrog, much like the *bletelach* discussed below.

2. Brown spots may develop from the esrog's being damaged *after* it was picked from the tree. Part of the peel comes off and the area turns brown. This has the same status as a black or white pasul spot.

If no other esrog is available, we may consider it a kosher spot, and the esrog can be used. If the damaged spot actually turned black, however, no such leniency exists.

If one is in doubt as to whether the brown spot was formed before the esrog was picked (where the esrog would be kosher) or afterwards (in which case it would be pasul), the esrog is considered kosher.

The rest of the types of brown spots all have the same status. Unless otherwise stated, they are kosher spots and, although there are those who invalidate the esrog where multiple discolorations develop, the custom is to always consider the esrog kosher.

3. Brown spots may be the result of damage caused by the sun. Much like human skin, the skin of the esrog becomes discolored when overexposed to the sun's rays.

4. Brown spots may also be the result of the secretion of the esrog's juice. Such spots will be discussed in more detail further on.

5. The strong ray of the sun sometimes burn an accumulation of dust on the esrog, thus causing a brown discoloration.

It should be noted that although the brown spot on the esrog may shrivel slightly in this scenario, the esrog is still complete and not considered damaged so as to render it pasul.

6. A brown spot may be the result of an insect's bite.

7. A brown spot may also be the result of an insect infestation. This will be described more fully later on.

8. Brown spots may be the result of dryness. We have already discussed the halachos of dryness in an esrog.

9. Pressure applied to the esrog may effect the secretion of liquid, causing an orange-brown stain. Although an esrog with such a stain is kosher as mentioned, it sometimes turns black, which would render it a pasul spot.

10. Green esrogim are sometimes subjected to ethylene gas in order to turn them yellow. An improper dose of ethylene may result in brown spots.

An esrog infested with insects.

Rebbe Mordechai of Neschiz zt"l lived in abject poverty. Over the course of the year he would save as much as he could, penny by penny, to be able to buy an esrog for Sukkos. Once, shortly before Sukkos, Rebbe Mordechai traveled to the city of Brod to purchase an esrog with the money he had painstakingly saved. As he walked along the road, he saw a fellow Jew sitting by the side of the road, softly crying to himself.

"Why are your crying?" he asked.

The man sighed and answered. "Rebbe, my horse is my sole means for making a living. Today my horse died! I no longer have a way to support myself and my family. What will I do?"

Without hesitation, the Rebbe took the purse containing the money he had saved and pressed it firmly into the man's hand. "Take this money," he said with a reassuring smile, "and buy yourself a new horse."

The Rebbe then returned home. When he reached his house, he announced joyously, "This Sukkos, while the entire Jewish people recite a blessing over their esrogim, I shall recite a blessing over 'a horse'!"

(Yeinah shel Torah)

Additional Halachos of Brown Spots

1. An esrog with a black spot on the *chotem*, that appears to be brown at first glance is kosher; if it has a brown spot that appears to be black at first glance, the esrog is pasul.

2. An esrog with a dark brown spot that tends towards black is pasul.

3. An esrog with one kosher brown spot and one pasul black spot that appear equidistant whichever way one views them is considered to be an esrog with multiple kosher spots.

4. An esrog with a collection of brown spots on its *chotem*, each of which, when viewed separately, does not tend to black, but together, when viewed collectively do tend to black, is considered pasul.

5. Red spots on the esrog are the subject of controversy. Some consider them pasul, like black or white spots. Others consider them to be like the brown spots discussed above, and therefore kosher.

An esrog with a red spot on its chotem.

Halachos of Kosher and Pasul Spots

1. An esrog with numerous kosher spots, is kosher but not *mehudar*. An esrog whose skin is a mixture of green and yellow is considered *mehudar*; these mixed colors are not considered spots but rather the natural color of the esrog's skin.[35]

---

35. However, there is an opinion that if the green shade of the esrog is dark, i.e. the color of grass, and parts of the esrog turn yellow, it is considered an esrog with numerous kosher spots and, therefore, not *mehudar* (Bikurei Ya'akov 648:44).

198 THE FOUR MINIM

2. A green spot is a kosher spot.[35a]

A green spot on the chotem.

35a. Heard from Rav Refoel Reichman.

3. Two or more different colors of pasul spots combine to render the esrog pasul, if found along the greater part of the esrog's circumference.

The Secretion of the Esrog's Juices

1. The secretion of the esrog's juices can result in three different discolorations of the esrog's skin:

   a. Ethylene is sometimes used to encourage the change of the esrog's color from green to yellow (this will be discussed in full later on). This may cause a dark green discoloration of the esrog. Such a spot is kosher, wherever it may occur. However, an abundance of such spots may cause the esrog to appear unsightly, and make it unsuitable for *hiddur mitzvah*.

   b. Bruises suffered by the esrog may also result in a brown discoloration. This discoloration is kosher, and its specifics are addressed above in the section discussing brown spots.

   This type of bruise can very often be caused by the constant handling and rubbing that result from the *na'anuim*, or shaking of the lulav.
   Interestingly, the Chasam Sofer writes that because these discolorations come about through the natural performance of the *mitzvah*, they actually make the esrog even more *mehudar*.

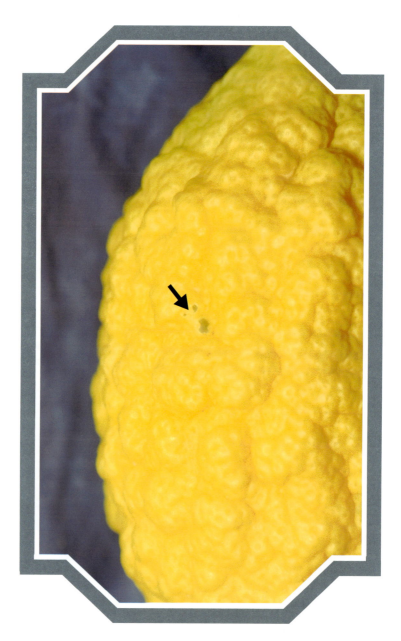

A green spot caused by exposure to ethylene.

c. If the esrog is held too tightly, this can cause its juices to seep inwards. This may result in a green tinge, distinct from the rest of the esrog. The green may later turn brown, which is kosher, or black, which renders the esrog pasul.

A brown spot caused by pressure.

2. This loss of the juice does not constitute damage that would render the esrog incomplete and pasul.[36]

3. In order to avoid the discoloration caused by secretion, it is advisable not to place the esrog on a hard surface such as a wooden table, but rather on the flax or soft-plastic casing that is used for packaging the esrog.

An esrog placed on a bed of flax.

---

36. Chazon Ish 147.

An esrog placed on the soft styrofoam plastic casing used for packaging.

Insect Infestations

1. Certain varieties of insects are known to infest citrus fruits, including the esrog. Among them is a small, white insect that spins a brown covering over itself on the esrog's surface. When sprayed with pesticides, the insect underneath dies, and the residue that remains appears as brown spots on the esrog's surface.

   The brown covering can be removed with a toothpick, wax, or modeling clay. Great care must be taken not to damage the esrog by scratching it or bruising it by pressing against it with unnecessary force.

   Beneath the brown covering, it is common to find a white mark. The mark may be either the remains of the insect, which would not effect the kashrus of the esrog, or a white discoloration that the insect caused to the esrog, which would be a pasul spot.

2. Dead insects found on the esrog are not considered discolorations. The insects remain external attachments to the esrog, and the Mishna Berurah clearly states that the issue of "kosher spots" applies only to discolorations of the esrog itself.

3. Even so, it is best to remove all remnants of the insects with modeling clay, in order to beautify the esrog. Furthermore, until the white marks are removed, one cannot tell for certain whether they are in fact insect remains that are kosher, or discolorations that are pasul.

4. Sometimes a black fungus mark develops beneath the brown covering. The fungus mark should be removable, unless part of the fungus has grown into the skin of the esrog. In such a case, although the mark is not removable, it is usually so small that it is unnoticeable, and therefore the esrog remains kosher, as discussed above.

Chotem

The *chotem* is the topmost area of the esrog, beginning where the sides of the esrog begin to slope inward and concluding with the esrog's tip. The Torah is much stricter regarding spots found on the *chotem* than it is regarding spots found on the esrog's sides. Spots on the sides will render the esrog pasul only if two or more of them cover the greater part of the esrog's circumference, while even the smallest noticeable spot found on the *chotem* will render the esrog pasul.

1. In regard to the color of the spots that can render an esrog pasul, it makes no difference whether they are found on the *chotem* or the sides. Black and white spots are pasul; brown spots are usually kosher; and red spots are the subject of debate among the *poskim*. The halachos of black, white, and brown spots have been discussed in greater detail above.

2. A pasul spot found on the *chotem* will invalidate the esrog only if it is noticeable at a superficial glance.

3. If an esrog has two slopes on the top half of the esrog, a bottom slope and a top slope with a narrow waist (having the appearance of a *gartel*, or belt) in between, both slopes carry the stringencies of the *chotem*. If the waist is sloped on one side, but straight on the other, the sloped side carries the stringencies of the *chotem* but the straight side does not.[37] Others disagree and always only consider the top slope as the *chotem*.[38]

---

37. Heard from Rav Chaim Kanievsky.
38. Heard from Rav Nissim Karelitz.

An esrog with a gartel.

4. An esrog with a bent *chotem* is kosher.

An esrog with a bent chotem.

The Origins of *Bletelach*

The outermost skin of the esrog is most sensitive to abrasion while the esrog is still attached to the tree, and within the first few days after it is picked. For this reason, it is very important to handle the esrog with care, and to place it only on soft surfaces such as the flax or soft plastic wrappers that are normally used in packaging.

Damage caused to the skin while the fruit is still attached to the tree can result in four different marks, all of which are grouped in the category of *bletelach*.

1. Some *bletelach* are caused by scratches made by the esrog tree's leaves on its tender skin. (These scratches are often so thin that they are not visible even under a magnifying glass.) The esrog then heals itself, forming scabs over the scratches, known as *bletelach*.

There is a common misconception that *bletelach* are leaves

The serrated esrog leaves can scratch the outer skin, resulting in marks known as bletelach.

Bletelach on an esrog caused by the serration of the leaves.

[*bletel* means "leaf" in Yiddish] that have stuck to the esrog during its growth. This is not the case, as we have explained.

2. Sometimes thorns cut the esrog's skin during its growth. The skin then heals itself by forming *bletelach*, as described above.

These thorns can even remove part of the esrog. In such a case, it is important to ascertain that the wounds have indeed fully healed, so that the esrog should not be considered damaged (incomplete), rendering it pasul.

An esrog damaged by thorns during its growth. The esrog healed itself by forming scabs known as bletelach.

3. A mark similar to *bletelach* can be caused by insects. The insects leave thin, white stringy excretions that can be scratched off with a toothpick or fingernail. Extreme care must be taken when removing these excretions not to damage the skin of the esrog.

Bletelach caused by insect excretions.

4. Another mark similar to bletelach is caused by insects that bite into the esrog. While healing itself the esrog produces a gray mark. Sometimes these gray marks appear as crevices in the esrog's skin, giving the appearance of a damaged, pasul esrog. However, since the esrog healed itself, these are considered kosher *bletelach*. Furthermore, the light shade of gray is not considered a pasul white spot; see our discussion on white spots.

The Halachos of *Bletelach*

1. *Bletelach* that did not fully heal before the fruit was picked theoretically render the esrog pasul. However, as explained above, damage that is not readily noticed by the naked eye does not render the esrog pasul.[39] Therefore, if the *bletel* appears to cover the entire scratch, even though under inspection with a magnifying glass a dull green mark is seen within the *bletel*, signifying that it has not fully healed, the esrog is still considered kosher.

2. *Bletelach* that are raised above the surface of the esrog to such an extent that they can be felt, are considered to be in the category of *chazazis*, and are a pasul discoloration. It is important to note that the coarse texture of the *bletelach* does not in itself render the esrog pasul. Therefore, if the *bletelach* can be felt because of their coarseness, but not their height, the esrog is kosher.[40]

In cases of great necessity, even an esrog with elevated *bletelach* may be used.

3. An esrog with *bletelach* that formed within a crevice on the esrog's skin, and rise above the crevice such that they are level with the surrounding skin, is considered kosher. Although such *bletelach* are indeed elevated, since they cannot be felt above the skin's surface, they are kosher.

4. Although some *poskim* forbid the use of esrogim with numerous kosher spots scattered across the esrog, as discussed above, *bletelach* do not fall into this category. Therefore, as long as the

---

39. Heard from Rav Nissim Karelitz.
40. Heard from Rav Nissim Karelitz.

elevation of the *bletelach* cannot be felt, they are kosher regardless of their number.[41]

5. Pale gray *bletelach* are kosher, as mentioned above in the discussion of white spots.

6. If one is comparing two esrogim, one of which has a beautiful shape but has some kosher *bletelach* that are slightly noticeable, and one of which is clean from *bletelach* but has an uncomely shape, it is preferable to choose the beautifully shaped esrog. The shape is more readily noticeable than the *bletelach*, and therefore since both are kosher, the more beautiful esrog should be chosen.[42]

> *"I shall be exalted through the date palm" (Shir HaShirim 7:9). The word for "I shall be exalted" is e'eleh, written with the Hebrew letters aleph, ayin, lamed, heh. Aleph is for esrog, ayin for aravah, lamed for lulav, and heh for hadas. This comes to teach us that the arba'ah minim should be bound together with the lulav, which grows from the date palm.*
>
> *(Likutei Torah, Volozhin)*

---

41. Heard from Rav Yosef Shalom Elyashiv.
42. Heard from Rav Ezriel Auerbach. However, such a decision is dependent upon the severity of the *bletelach* or the misshapenness of the esrog.

A Pasul Spot that was Removed

The status of an esrog invalidated due to pasul spots that were then removed in a way that did not damage the esrog itself, is the subject of debate among the *poskim*. The Mechaber permits its use, and holds that a *berachah* may be recited over it even on the first day of Yom Tov. The Rema, however, holds that it is pasul for the entire duration of Sukkos. In cases of great necessity, it may be used for the rest of the days of Sukkos, but not for the first day of Yom Tov.[43]

Although the Sephardim customarily abide by the ruling of the Mechaber, since this is a case of *safek berachos lehakel* (in cases of halachic uncertainty a *berachah* may not be recited), the Kaf HaChaim rules that Sephardim as well should preferably not use such an esrog.

Removing the elevated *bletelach* from an esrog, however, renders it kosher according to all opinions.

Soaked or Cooked Esrogim

1. A cooked esrog is pasul.

2. An esrog that was soaked in water for twenty-four consecutive hours is considered as if it was cooked, and is therefore pasul.

3. An esrog that was soaked in vinegar or any other pungent substance for the amount of time that it would take to bring the vinegar to a boil is pasul.

---

43. Shulchan Aruch 649:5; Mishnah Berurah s.k. # 38.

A Black or Orange Esrog

1. A truly black esrog is pasul, even in places where such esrogim are commonly grown. If the esrog only tends to black, but is really a darkish green, it is kosher only in places where such esrogim are commonly grown. In other places, even those that are adjacent, such an esrog is pasul.

A black esrog.

2. An esrog with a deep orange color, bordering on gold, is kosher. However, there are those who feel that such an esrog is not beautiful, and therefore prefer not to use it.

A golden-orange esrog.

A Long and Thin Esrog

A long and thin esrog that satisfies the minimum size requirement, is kosher.[44]

A long and narrow esrog.

---

44. *Kashrus Arba'as HaMinim*, page 10.

A Perfectly Round Esrog

1. An esrog that is as round as a ball is pasul.

An esrog round as a ball.

2. A round esrog with a *pitam* is kosher, but only if the esrog is not perfectly round. Others do not make such a distinction, and hold that a *pitam* validates any round esrog.[45]

---

45. Heard from Rav Nissim Karelitz.

An esrog that is not perfectly round and also has a pitam.

A round esrog with a pitam.

3. An esrog without a *pitam* whose body is round except for the tip of the *chotem*, which juts out slightly, is kosher.⁴⁶

An esrog that is perfectly round, except for the tip of the chotem which juts out slightly.

---

46. Heard from Rav Nissim Karelitz.

> *Every single Jew has within himself attributes and characteristics that are represented by the Four Species. A person may excel in certain areas, while being faulty in others. To the extent that a person uses his positive traits to the good of the entire Jewish people, he receives in return the opportunity to correct the traits in which he is lacking.*
>
> *(Sefas Emes, 5636)*

An Esrog that was Grown in a Mold

If a mold was placed over an esrog while it was still attached to the tree, such that it grew to fit the shape of the mold, the esrog might be pasul, depending on its shape. If the esrog developed a shape unusual to the species, it is pasul. If it developed a shape similar to what is common to the species, it is kosher. This even includes the shape of a water wheel, i.e. with spokes jutting from its sides, as being kosher.

Twin Esrogim

Two esrogim that have grown attached to one another at the stem, and separate at the top, are of questionable kashrus. Therefore it is best not to use them; however, in cases of great necessity it may be used.

Twin esrogim.

A Straight-Sided Esrog

1. It is better to be stringent and not use an esrog with straight sides, instead of the standard curved shape. However, *b'dieved* such an esrog is kosher.[47]

A straight-sided esrog. It is best not to use such an esrog.

---

47. *Kashrus Arba'as HaMinim*, page 61.

2. A straight-sided esrog with a *pitam* is kosher.[48]

A straight-sided esrog with a pitam.

---

48. Heard from Rav Nissim Karelitz.

A Green Esrog

1. A dark green esrog, the color of grass, is pasul. However, if the esrog will turn yellow over time, it is kosher even before it turns yellow.

A dark green esrog that borders on black and will not turn yellow.

2. There are those who hold that the esrog must be mostly or entirely yellow; however, the accepted ruling does not follow this opinion as we just stated.

3. The Mishnah Berurah states that *l'chat'chilah* it is best to use an esrog that has at least begun to turn yellow, in order to verify

that it is not the pasul, dark green esrog discussed above. However, experience has proven that all of the green esrogim that are marketed today do turn yellow over time. Therefore, Rav Yosef Shalom Elyashiv permits the use of green esrogim even though they have not yet begun to turn yellow.[49]

A dark green esrog that can turn yellow.

---

49. Heard from Rav Ezriel Auerbach.

The Gemara states that of the Four Species, two of them are associated with fruit (the date palm and the esrog), and two of them are not (the hadas and the aravah). One cannot fulfill his obligation until he binds all four together. So too, the Jewish people cannot be answered in their prayers in times of need, unless the entire spectrum of the community is united.

The species that bear fruit represent the Torah scholars, who promulgate Torah study among the Jewish people, and pray for the welfare of the other Jews, that they should be spared from all distress. Similarly, our Sages teach us that the grapes of the vine should pray for the welfare of its leaves. If not for the leaves that protect the vine, the grapes could never survive. The grapes represent the Torah scholars, and the leaves represent the common Jews. If not for their support, the Torah scholars could never survive.

Thus, each group benefits and complements the other.

(HaManhig)

Artificially Stimulating the Color Change of an Esrog

A green esrog can be stimulated to turn yellow by exposing it to a chemical called ethylene. Ethylene is found in aromatic apples of all varieties. A simple way to stimulate the color of a esrog is to place it together with an apple in a sealed container. The esrog should first be taken out of any plastic bags encasing it,

Yellowing an esrog by use of an apple.

but should be left wrapped in its plastic styrofoam package to prevent damage. The esrog then absorbs the ethylene gas emitted by the apple, and turns yellows. The container must remain closed for the duration of the yellowing process in order to keep the ethylene from escaping.

It is important not to expose the esrog to ethylene for too long, because this may cause the stem to fall off, either during the yellowing process or later. The correct amount of time for achieving best results while preserving the stem depends upon the temperature and humidity. In Israel, in a relatively cool, dry climate such as Yerushalayim's, the esrog should be left exposed to the apple for two or three days. In the hot and humid climate of Bnei Brak, a day and a half will suffice.

The esrog should be exposed to ethylene only in the period immediately preceding Sukkos. If the esrog is exposed to ethylene more than a month before Sukkos, it may develop a golden tinge. Although kosher, gold is not considered a beautiful color for an esrog.

If for some reason it is necessary to turn the esrog yellow long before Sukkos, it is best to put it aside with the apple for no longer than eight hours. Thus the ethylene will only begin the yellowing process, and the esrog will continue to turn yellow naturally in the following weeks. This method may cause certain spots on the esrog turn to yellow, while the rest remains green. Although such an esrog is kosher, it may not be considered a beautiful esrog.

An esrog that is partially yellow and partially green.

The numerical value of the Hebrew word esrog is 610. If we add 3, representing the other three species, the total becomes 613. Thus, one who fulfills the mitzvah of arba'ah minim is considered as if he upheld all 613 mitzvos.

(Abudraham)

After their forty-year trek through the barren desert, the Jewish people rejoiced to arrive at Eretz Yisrael, a bountiful, lush and beautiful land. As a remembrance of their joy and gratitude to the Almighty, they were commanded to take the arba'ah minim. These were a sampling of the best among the plant world.

The arba'ah minim had the following advantages as well:

1. *They were plentiful in Eretz Yisroel at that time, and therefore it was easy for every Jew to fulfill the mitzvah.*
2. *They are beautiful and fresh in their appearance. The hadas and esrog have a pleasant fragrance, as well. The lulav and aravah, although not fragrant, do not have a foul odor either.*
3. *They maintain their freshness. Other fruits such as pears would whither or rot over the course of Sukkos.*

(Moreh Nevuchim 3:43)

Another way of exposing the esrog to ethylene is to wrap an apple in flax, thus allowing the flax to absorb the ethylene from the apple. When the esrog is later wrapped in the same flax, it absorbs the ethylene and turns yellow.

On an industrial scale, there are some esrog growers who store their picked fruit in sealed rooms with piped-in ethylene gas, thus enabling them to make hundreds of esrogim yellow at once.

Esrogim being exposed to ethylene in an industrial gas room.

An Esrog that was Placed beneath a Bed

An esrog must not be placed beneath a bed. The halacha states that an impure spirit rests beneath beds, making it dangerous to eat any food placed there.[50] Since one of the conditions of a kosher esrog is that it must be edible, an esrog that was placed under a bed is considered pasul.

If, *b'dieved*, it was placed there, and no other esrog is available, it may be used but it is questionable whether a *berachah* may be recited.[51]

This halacha applies both by day and by night, and it is irrelevant whether or not someone was sleeping on the bed at the time.

The above applies only to the first day of Yom Tov. For the rest of the days of Sukkos, an esrog that was placed under the bed is kosher.[52]

> *One must pray fervently in order to merit a mehudar esrog. He should cry out to Hashem on Rosh Hashana and Yom Kippur to grant him one. We have no idea how precious and significant is this mitzvah. Our minds cannot possibly comprehend its great holiness. Through it, we hasten the building of the Beis Hamikdash, may it be speedily in our days, amen.*
>
> *(Likutei Eitzos, Rabbi Nachman of Breslav)*

---

50. Shulchan Aruch, Yoreh De'ah 116:5.

51. Kaf HaChaim 649:80. The Steipler Gaon ruled that a *berachah* should not be recited. The Chazon Ish and Rav Yosef Shalom Elyashiv rule that a *berachah* may be recited.

52. Shulchan Aruch 649:5; Mishnah Berurah s.k. # 45.

Deep, heartfelt thanks
to my dear and close friend,
a person who is constantly
pursuing opportunities to perform
acts of charity and kindness,
joyfully

Mr. Edi Betesh הי"ו

In the merit of his generous support
of this sefer,
may Hashem's blessing
of success
be granted to him,
his entire family
and to
מרן ראש הישיבה הגאון
Rav Yehuda Adess שליט"א
Rosh Yeshiva of Kol Yaakov.
May the Rosh Yeshiva be granted good health,
both in body and spirit,
prosperity and happiness
all the days of his life, Amen

In memory of

a cherished man
who served the community faithfully —
a *shochet, chazzan,* and *mohel*

Reverend Mendel Klein ז"ל

who passed on to his eternal reward on
20 Adar 5748

ת.נ.צ.ב.ה.

Dedicated by
Shea and Malka Klein
Nuchem and Leah Sauber
Chaim and Charna Klein
Donald and Yocheved Liss

לזכרון עולם בהיכל ה'

In memory of a cherished man,
who constantly involved himself
in tzedakah and chesed
in every way

Mr. Nachman Leib ben Avraham ז"ל

who passed on to his eternal reward
4 Kislev 5764
ת.נ.צ.ב.ה.

Dedicated by his family
Mr. Yaakov Chaimovici הי"ו
Mr. Yehoshua and Mrs. Chana Devorah
Chaimovici הי"ו
Mr. Shlomo and Yael Dworkin הי"ו
and their mother Adel תחי'

May it be the will of Hashem
that they merit Jewish *nachas*
from all their children
and that they live to welcome
the coming of *Mashiach* in the Holy Land,
speedily in our days, Amen.